# Is Evolution Compatible with Christianity?

# Is Evolution Compatible with Christianity?

by
Christopher Gieschen

WIPF & STOCK · Eugene, Oregon

IS EVOLUTION COMPATIBLE WITH CHRISTIANITY?

Copyright © 2019 Christopher Gieschen. All rights reserved. Except for brief quotations in critical publications or reviews, no part of this book may be reproduced in any manner without prior written permission from the publisher. Write: Permissions, Wipf and Stock Publishers, 199 W. 8th Ave., Suite 3, Eugene, OR 97401.

Wipf & Stock
An Imprint of Wipf and Stock Publishers
199 W. 8th Ave., Suite 3
Eugene, OR 97401

www.wipfandstock.com

PAPERBACK ISBN: 978-1-5326-5703-0
HARDCOVER ISBN: 978-1-5326-5704-7
EBOOK ISBN: 978-1-5326-5705-4

Manufactured in the U.S.A.                                                 10/07/19

# Contents

*Acknowledgments* | vii

Chapter 1
In the Beginning: How I Got Started | 1

Chapter 2
Does Neutral Ground Exist? | 8

Chapter 3
Definitions or What Does This Mean? | 21

Chapter 4
Christianity: The Bible's or Dr. Kenneth Miller's? | 30

Chapter 5
Is Theistic Evolution a Valid Compromise? | 42

Chapter 6
Philosophy 101 | 50

Chapter 7
Darwin Right and Darwin Wrong | 62

Chapter 8
Time is of the Essence | 73

Chapter 9
Fossils as Evidence for Which Side? | 83

Chapter 10
Textbook Evolution Evidence Explained | 96

Chapter 11
The Origins of Everything | 109

Chapter 12
Stephen Meyer Weighs in and You Should Too | 119

Chapter 13
Has Alien Life Evolved? | 130

Chapter 14
The Word of God and This Issue | 140

Chapter 15
Four Websites Worth Your Time | 152

Appendix: Study Questions and Answers | 155

*Bibliography* | 163
*Index* | 169

# Acknowledgments

First I would like to thank Elizabeth Hoham, who teaches English classes at Concordia Lutheran High School (CLHS) in Fort Wayne, Indiana. She took on the job of preparing my bibliography for this book. . .a not so easy task indeed. Ever-thorough, she saw it through to the end. Hats off to her!

Next on my list is Jackie Gudel, wife of my best friend at CLHS, Rev. Joe Gudel. After finding out that paying for the rights to use actual biology textbook illustrations would cost more than I was willing to pay, she said, "Let me give it a whirl." Soon she finished one test illustration to see if it fit my needs. I could not have asked for better and the price was right for the rest of the illustrations. Look for the list of credits to see which ones are hers.

I would be remiss in not thanking my own adult children: Kevin, Jamie, and Joy for all their prayers and support in completing this project. They knew full well that I taught the material contained herein in my biology classes at CLHS. After having their backs all those years before, it was nice for them to have my back on this book, especially after it had been turned down prior to Wipf and Stock giving me their thumbs up.

I save the best for last. My wife, Leann Gieschen, has been instrumental in encouraging and all-around supporting in various ways to make this book become a reality. What many do not know is that she purchased my first creation apologetics book back in 1981, *Evolution: Possible Or Impossible* by James F. Coppedge. I had always known that God was the Creator and that evolution in the macro sense was totally wrong. I did not know how to show it so. After devouring that book she always gifted me with more. That translated into more effective creation apologetics in my science classes I taught. Leann, I love you for all you have done.

## Acknowledgments

The figures in this book are provided with permission from the following:

Answers in Genesis – Figures 1–4, 6, 7, 15

Master Books – Figures 7

Icons of Evolution – Figures 8, 13, 14

Jackie Gudel (illustrator) – Figures 5, 9a–9c, 10–12, 17

Public Domain – Figures 16a–16c

Copyright Clearance Center for book: Organization in Vision: Essays on Gestalt Perception – Figure 18 (Figures 19 – 21 based upon this figure)

## Chapter 1

## In the Beginning
### How I Got Started

"You never learn that the gravest issues may depend upon the smallest things."
—Sherlock Holmes, "The Adventure of the Speckled Band"

As Holmes explains to Watson, the smallest incidents or facts can have huge significance when it comes to solving a case. In other words, to Holmes, there really is no such thing as a useless fact. He seems to be admonishing Watson about being too stubborn when he complains that Watson "never" learns this truth. Yet anyone who has read *Sherlock Holmes* may recall several instances when Watson actually does try to carry on investigations the "Holmes" way, looking at *all* the facts, no matter how big or small.

It is the same with the creation versus evolution issue. To begin examining the topic, it is important to understand that there are three factors influencing a person's worldview: (1) religion, (2) upbringing, and (3) life experiences. Each area is a significant factor for whether a person becomes a believer of either creation or evolution. But, as the Holmes quote suggests, each of these areas may be impacted by a few seemingly small occurrences. This holds true for how I became a biblical creationist. There were many small events in my life that led me to hold creation as true. But first, let me talk about Joe.

Let's say Joe decides to be a Christian missionary. How does he come to that decision? Mainly, Joe's religion specifically tells him to do so. Jesus' instructions were written down by his followers and passed down to Joe via the Bible. For example, Joe can cite Matthew 28:19–20, where Jesus said, "Go therefore and make disciples of all nations, baptizing them in the name of the Father and of the Son and of the Holy Spirit, teaching them to observe all that I have commanded you. And behold, I am with you always, to the end of the age." Note that Jesus instructs his followers to spread the gospel to the rest of the world. Christians like Joe are to obey all that Jesus has commanded. Each person is to spread the gospel to others, who then become followers of Jesus, who then tell others, who then become followers . . . and you get the idea.

Are there other Bible passages Joe could point to that tell him to go and do mission work? Indeed, there are. Acts 22:15 says, "For you will be a witness for him to everyone of what you have seen and heard." Another is 1 Peter 3:15: "But in your hearts honor Christ the Lord as holy, always being prepared to make a defense to anyone who asks you for a reason for the hope that is in you; yet do it with gentleness and respect." We can see that Joe's religion is a compelling factor in his decision to become a missionary.

The second influence on Joe's decision to be a Christian missionary is his upbringing. Both of Joe's parents are Christians who spoke about Jesus in their home. They talked about mission work during family devotions, and his parents often financially supported missionaries. Joe's mother told him how a person had once shared the gospel with her when she was younger and led her to the Savior. Joe's parents raised him to see the world's treasures as blessings to be used for God's kingdom instead of being wasted on his own pleasures or wants.

As you can see, the way Joe was raised plays a part in his worldview formation. It also is true that the Holy Spirit plays a large role in this, but to Joe, it is his parents that have a significant role here. It is important to note that Joe's parents not only talk the walk but also actually walk the walk. Proper modeling goes a long way to fostering proper worldview growth.

Finally, life experiences can have a huge impact on a person's worldview. Joe's story is no different. His church sponsored missionaries, who came and talked about their work for the Lord in foreign lands. This increased Joe's interest in seeing how other cultures lived. He also read accounts of people who sometimes risked their very lives to bring the good news to others. These brave ones inspired Joe. During college, he was given the chance to work as an intern for a Bible society one summer and even spoke directly to a few missionaries who had returned home to visit relatives in the United States.

Life experiences are most critical in the later years prior to adulthood. This is why many churches wisely include a variety of experiences for their teens to help shape their worldviews. True enough that most will not become missionaries for Jesus like Joe. However, many will be impacted to become a Christian who shares his/her faith easily with others.

These three factors greatly influenced Joe in his decision to enter the mission field. As you can see, some of the causes were seemingly small, but others were more direct in their impact. But each was used to shape Joe into the person he came to be. My story is no different, but I will start with one particular memory of my upbringing.

From my earliest days, I can remember my parents taking me to worship at a Lutheran church. Paying attention and not disturbing others during the service was very much instilled in my brothers and me via my mother, as my father was busy serving as an organist at another church in downtown Chicago. All it took was a pinch on the leg and the offender rapidly snapped to attention. I never could figure out how she did it while looking straight ahead. As you can see, those little pinches (upbringing) reminded me to take what was going on in church seriously!

My Christian faith, under the guiding hand of the Holy Spirit, grew due to several factors. I recall having theological discussions with my dad over Sunday dinner or during our drives back from my grandparents' house. Dad was patient, letting me ask question after question, and he explained God's truth in ways I could understand.

Another influential person in the development of my belief system was my religion professor at Concordia Teachers College in River Forest, Illinois (now Concordia University Chicago). I will never forget his opening question to our class, most of whom considered themselves to be Christians: "Why are you a Christian?" Most attributed our choice in religion to our parents who dragged us to church when we were young, or that we were Christians because our parents were. His response brought us up short: "Then I am afraid that none of you are saved." After our shock and outrage died down, he challenged us further: "Care to show me from the Bible where it states that you are saved by your parents' faith or by them dragging you church?" His question made me recall the time when the Jews argued with Jesus about this very idea. They said they were saved because Abraham was their father (John 8:31–38). Needless to say, we were not successful in using the Jews' argument either.

This same professor also helped me deal with a time in my life when I questioned my faith. I asked him how I could be sure Christianity is the correct religion. After all, there are so many religions in the world, how can I be sure that it is the one that will save my soul? He encouraged me to do some research and find out for myself. Therefore, keeping one foot firmly planted on the rock, I dipped my other toes into a sea of other religions. I basically found that all other religions are full of dos and don'ts with no real guarantees. Christianity is the only belief system that involves God doing everything for a person, including sending his only Son, Jesus, to die for his lost creation and rise again. No other belief system has a god who would (or could) do such a thing.

But to bring us back to the topic of this book, I'll share how life experiences led me to where I am now. One Christmas, I received a dinosaur play set. It included several prehistoric-looking trees and plants as well as a big plastic rock for the creatures to climb. Interestingly, it also contained six "cave people" in various poses, yet there were no prehistoric mammals. Obviously at the time, I did not see the "error" that any evolutionist would. Today, one can still find those types of play sets with cavemen and dinosaurs, but now Ice Age mammals are thrown in. One wonders why the evolutionists do not publicly bemoan the toy industry supporting creationist "nonsense" of men and dinosaurs existing at the same time. One also wonders whether the toy industry even realizes it is supporting biblical truth found in Genesis. But this play set began my love of things prehistoric.

After earning my teaching degree I was hired by a Lutheran high school to teach biology and english. It was there where I became exposed to the creation and evolution debate. I do have an early teaching memory of a parent questioning me during a "back to school night" session. She wondered if I taught evolution in my biology classroom. I answered that I did, and that I also stressed to my students that what the Bible says about creation is true. At that time, I did not realize just how ineffective my teaching style was in training future apologists.

While attending science conferences, I came to see that many of my public school colleagues often wondered how I compartmentalized my brain in holding two seemingly contradictory beliefs as being true: evolution and creation. For many teachers and scientists, there is not even one sliver of compromise or relationship between the two. In their minds, one cannot exist if the other does. To many, the Bible is just a bunch of myths borrowed from various pagan religions that were twisted to the Hebrew concepts of monotheism. People who hold such a belief sometimes get upset when I say that I look at them as both/and, not either/or as being true. What I mean by either/or is that some Christians deny everything about evolution because it's not mentioned anywhere in the Bible. I fully understand this, but it is misguided due to not having a correct definition of evolution. The problem with evolution's definition is that it has now become so generic. Some say it means we all came from one ancestral cell, which is not a scientific fact at all (Evolution Wrong). But others, including myself, say it means that we have variations within specific species, such as the different breeds of dogs or cows. This is absolutely true in every sense, and is what we call microevolution (Evolution Right).

So in a very limited sense, I do believe in evolution, but not how the evolutionists would like me to believe it. I also hold fast to the word of God as found in Genesis all the way to Revelation. When God said he created the world by his word, this is exactly what happened. He most certainly did *not* use evolution to create the world. Had he done so, he would have said so. Later on, we shall explore the errors made when trying to merge biblical truth with secular science.

Another event that led me to become a creation apologist occurred at a National Association of Biology Teachers conference in the early 1980s. This conference took place around the time when a small group in California known as the Institute for Creation Research (ICR) was voicing the first serious and scientific objections against evolution. I was amazed to see such vitriol and passion in some of the knee-jerk reactions against that early creationist group. At one of the sessions I attended, I asked the presenter if it were possible to present both sides and let the students make up their own minds. He responded testily, "Of course not! Do we let them decide if two plus two equals four?"

I wish I had known how to respond to him at the time. Had I known then what I know now, I would have told him that comparing evolution to math is comparing apples to oranges. I can prove that two plus two is four. No one can scientifically prove particles became people.

After that conference, a craving for knowledge that upheld creation took over my mind. My wife gave me my first creationist book, *Evolution:*

# Is Evolution Compatible with Christianity?

*Possible or Impossible* by Dr. James F. Coppedge. The subtitle intrigued me even more: *Molecular Biology and the Laws of Chance in Nontechnical Language*. This book opened up a whole new world of statistics and scientific facts that confirmed the impossibility of evolution creating life. Details of the cell's intricacies—namely those concerning DNA, RNA, and protein synthesis—were explained. The author's premise was that there is no possible way information-rich molecules could have put themselves together. The odds against it are just too great. Nevertheless, famous evolutionists, such as Stephen Jay Gould, speak about "fortuitous events," which can circumvent such bleak odds.[1] Coppedge's book turns the chance events that Gould says make evolution possible into simple flights of fancy. After reading what Coppedge has to say, it's easier to believe in fairy tales than the ideas of organic evolution, which supposedly explains where the first cell came from.

Coppedge's book foreshadowed the excellent work of Stephen C. Meyer, who wrote *Signature in the Cell*. To make an excellent but thick book short, it is DNA that is the signature of the intelligent agent who designed the first cell. Coppedge stated that DNA is a code molecule that carries a message.[2] We all know that letters, such as those forming these sentences, are elements of a code. Messages convey information. Meyer's work completes Coppedge's idea that codes, like DNA, cannot form themselves. They need an agent, and an intelligent one at that.

A word of caution is needed when one purchases a book dealing with this subject. Peruse it prior to purchasing! Another book my wife purchased for me was entitled *God Did It, But How?* by Robert B. Fischer. She thought it sounded excellent based upon the title. Imagine her shock when I told her she had purchased a theistic evolution book. Theistic evolutionists hold that God used evolution to do his handiwork. I told her not to worry, as I had to learn how to refute these erroneous ideas as well as those of materialist evolutionists—those who believe that the "trinity" of time, matter, and chance is the creator of everything that exists.

After some years of reading various creation publications, books, and pamphlets, I began to include in my science lectures the facts and interpretations I had learned. I also began to lead a few Bible studies at church, showing how we can trust that God's word is true, as it says it is. Once the internet flourished, a new source of science supporting creation was

---

1. Gould, "Evolution of Life."
2. Coppedge, *Evolution*, 137.

found. I next tried to share these revelations with science teachers at our state's science convention. As I figured, there was only a handful that supported creation. Needless to say, as of the writing of this book, I am presently "expelled" from presenting at that convention. I have even offered to teach a session on critical thinking, but apparently the conference program committee prefers students to not think critically about evolution, stem cell therapies, or other controversial biology-related topics.

Throughout my years of teaching high school science, my students have expressed their doubts on whether they could handle a discussion of creation versus evolution—they doubt their ability to defend their beliefs. They claim that they could never discuss the issue as well as I could. When I ask why they think I'm so much better at it, they reply, "Because you're so smart." But here's the truth: I am just better read on the subject. All it takes is immersing one's self in books, articles, and websites wisely; role-playing in mock discussions among peers or with friends who do not hold any creation convictions can also be extremely helpful.

So now we come to the writing of this book—a book for those who seek to better handle one of the major stumbling blocks people face when it comes to believing in Jesus as their Lord and Savior. I also wrote it for those who have questions or doubts about evolution being all that it claims to be. Such readers will find out that their doubts are well founded. I will also pose questions that might come across your mind in italics. The answer to each will follow.

Finally, I wrote this book for my students, who usually never save any of their notes from class, so that they will have something useful in hand when they go to college. Perhaps they can lend it to roommates or classmates when dealing with this topic in the university setting. But no matter why you picked up this book to read, I pray that God will bless your time spent perusing it, and that you may find it helpful in discussing the issue of evolution versus creation with those around you.

# Chapter 2

# Does Neutral Ground Exist?

> "There is only one point on which I must insist. You must not interfere, come what may. You understand?"
> "I am to be neutral?"
> —Sherlock Holmes and Watson, "A Scandal in Bohemia"

In "A Scandal in Bohemia," Sherlock Holmes basically tells Watson to follow his instructions exactly. Watson is not to become involved with the plan Holmes has in mind. So Watson asks the question above. Therein hangs the point of this chapter. Neutral . . .

*What exactly does the word neutral mean?*

A number of definitions come to mind. Neutral can refer to a color that is rather dull and lifeless, such as beige. It can mean the car gear between reverse and drive. Or it just might mean belonging to neither one of two sides. Think of neutral as being a buffer zone between two groups; it gives no advantage to either side.

*What does neutral have to do with creation and evolution?*

In a very well done video, *The Ultimate Proof of Creation*, Dr. Jason Lisle explains the importance of "neutral" in the creation versus evolution discussion. When you begin discussing creation with an evolutionist—perhaps

## Does Neutral Ground Exist?

your teacher, college professor, or friend—the person with whom you are speaking may soon discover that you are a biblical creationist. So he or she will ask you nicely if you could agree to leave your presuppositions behind and discuss the issue on "neutral ground." This neutral ground is science.

At first thought, this seems fair to you. After all, science is generally not openly hostile to the Bible. For example, you know that living creatures are made up of cells, and so does the scientist. The Bible makes no claims about this, so it seems like discussing creation and evolution on neutral ground is a reasonable request. However, please be aware that doing so will set you at a disadvantage. The evolutionist is asking you to give up your entire belief system, while he or she gets to retain his or hers. It is not really science that the evolutionist wants to discuss. The truth is that he or she is standing upon one of two belief systems—both of which are anti-Bible. The first belief system is *methodological naturalism*.

### What is methodological naturalism?

Methodological naturalism claims that this world and space (galaxy and universe) is all that there is.[1] Nothing else exists except nature. Now perhaps you can understand why it is called *naturalism*. A perfect synonym for this is materialism. Materialism is the belief that matter is the only reality that exists. A materialist will seek to explain everything as a result of interactions between all the forms of matter and energy that exist.

Next, it is important to know that methodological naturalism is part of the scientific method. This is where the first part of the term comes from. It also has a permission clause with it. This clause states that scientists are able to believe in anything else they wish to believe in. So belief in angels or even God is allowed. But said scientists must act as if naturalism is true and the only reality that exists. Thus they cannot invoke God or any supernatural agent as a cause for anything. This is fine and good for most of what science does for mankind such as find more energy-efficient machines or cures for cancer. However, this philosophical construct is powerless to explain things that are what I will call *singularities*—events that only happen once and never again. For example, if we are talking about the origin of the universe or the creation of life, these events have only happened once and no human was there to see them happen. Therefore, any study of these events is not amenable to the usual methods used by some scientists.

---

1. Commission on Theology, *In Christ*, 81.

# Is Evolution Compatible with Christianity?

Think about what it's like to look for clues at a crime scene. After reviewing the scene, a trained detective ought to be able to figure out what happened. In addition, his or her explanation must be logical and must not violate known scientific laws. This is where evolutionary scientists get into trouble. The quandary they find themselves in is dealt with in greater detail in chapter 11. If life does indeed have a supernatural origin, then scientists are wasting resources and money trying to find a natural explanation for that origin. Think about it. Whether we come from the hand of God or some other super intelligence, or whether we come from some incredibly fortunate series of accidents, does it really matter when it comes to using science to ease suffering or make life better for the impoverished nations? No.

*What is the second belief system a scientist may hold?*

*Scientism* is a belief system that does not simply claim that science is a good and trusted way to learn about nature (which, by the way, no one disputes); rather, it states that a particular approach to science, namely materialistic science, is the only way to gain knowledge when investigating nature.[2] So while you believe that God made all things in seven days as Genesis teaches, the evolutionist will state that science has proven this false. We will explore this more in depth in a later chapter, but for now, please be assured that science, as most of us understand it to work, has done nothing of the kind.

*Isn't it true that you don't need to believe in evolution to be a scientist?*

This is most certainly true, but this does not help with the problem that most Christians have surrendered their belief system or have a special box in their brain in which to place their worldview. It comes out on Sunday or perhaps during a special time of stress or pain, but when all is right with the world, it goes back into the box. This double thinking pattern was proposed by Stephen Jay Gould, and he called it *nonoverlapping magisteria* (NOMA).[3] This very idea seems to recall Galileo's remarks when the church, (really the scientists of his day and not the church) was at odds with his statements about planets orbiting the sun. Galileo is reported to have said that religion tells us how to get to heaven, while science tells us how the heavens operate. What I find most interesting is that the evolutionist forgets Galileo was talking about two different heavens, for heaven's sake!

---

2. Commission on Theology, *In Christ*, 5.
3. See Gould, "Nonoverlapping Magisteria," 16–22.

## Does Neutral Ground Exist?

The first was where all believers in Jesus will end up, and the second was the realm of the planets and stars.

For the most part, when any scientist, Christian or not, conducts an experiment, makes observations, or does any of the activities normally associated with science, he or she does not need the Bible or any other nonscientific source. The same pertains to evolutionary thinking, so swap evolution for "the Bible" in the previous sentence to see the truth. The real truth is that most work in any scientific discipline does not require belief in evolution.

*Then why does the evolutionist ask to remove the Bible from the discussion?*

Understand what was said in chapter 1. Everyday science has really nothing to say concerning the issue of origins. All we have are facts. What the facts can tell us rests squarely upon our presuppositions or worldviews. It is like looking at anything outside on a sunny day with two different pairs of sunglasses. Use the blue tinted ones, and everything is tinged blue. Use the dark green ones, and to no surprise, everything seen is hued green. This is why Christians see the world and the facts in it so differently from non-Christian scientists or college professors. The danger comes when Christians are tempted to see science as the end-all when it comes to anything one can say about humans.

Are we special creations of a Supreme Intellect who loves us more than we can fathom? The evolutionist would answer, "No." Is there such a thing as sin? Again, they would say no. Does the soul exist? "No," the evolutionist replies. "Well, that is okay as long as I have Jesus," a Christian might respond. But science can even cause Christians to doubt Jesus too. After all, according to secular science, people cannot rise from the dead, so the resurrection must never have happened. Hopefully you can see that this implication is very serious indeed.

What is so ironic about all this is that most secular scientists have forgotten it was Bible-believing Christians who even made science possible. Galileo said, "Whatever we read in that book [nature] is the creation of the omnipotent Craftsman."[4] This is but one example of the many quotes you will find the founding fathers of science to have said and believed. The Bible in no way hindered their research and findings.

Let us consider how we are to approach the statements made by evolutionists about origins. Think for a moment that the universe we inhabit

---

4. Galilei, *Dialogue*, 3.

## Is Evolution Compatible with Christianity?

is the result not of God saying, "Let there be," but rather the product of a huge explosion. Have you ever seen explosions build more complex items? Science also claims that we all are really just animals anyway. Before you disagree with this, know that Peter Singer, a Darwinian materialist, said that it is "speciesism" (like racism or sexism) to suppose there is anything special or unique about human beings.[5]

Continue pondering the notion that nature can explain things as they are. Is it reasonable to say that reason ought to exist? If you search one of the premier creationist resources, the Answers in Genesis website, you will find a more complete discussion by Dr. Jason Lisle on this very concept of reason being reasonable only if creation as stated in Genesis is true.[6] But there is more to ask. Evolution says that *mutation* is the instigator of change. A mutation is any change in the DNA code for a protein. According to evolution, humans came from fish. Can we really trust our thoughts if we have mutated fish brains? Think about it! Oh. . .wait a minute. . .(Consider this previous sentence very carefully. . .you have to use a mutated fish brain. . .)

Another disquieting notion regarding giving up a biblical worldview and replacing it with a purely godless evolutionary one is revealed in some comments made by Bill Nye, famous for popularizing science with younger children. As you read the following comments from Nye, compare what he says with what you know the Bible tells you about yourself.

> I'm insignificant. . . . I am just another speck of sand. And the earth really in the cosmic scheme of things is another speck. And the sun an unremarkable star. . . . And the galaxy is a speck. I'm a speck on a speck orbiting a speck among other specks among still other specks in the middle of specklessness. I suck.[7]

As you can see, Nye's words hardly agree with the Bible, which declares that those who believe in Jesus are sons and daughters of the most high. We can even call the Maker of heaven and earth a very intimate title: Abba or Daddy.

Later on in the same speech, Nye confuses the two types of science that are discussed later on in this chapter. They are operational science and

---

5. Singer, *Animal Liberation*.

6. Jason Lisle, "Ultimate Proof for Creation/God's Existence," https://youtu.be/-j9-cyRbFcs.

7. Quoted by Luskin, "Real Science."

origins science. He ought to know full well that one does not need to be an evolutionist to improve our technology. Yet read what he says:

> Our understanding of evolution came to us by exactly the same method of scientific discovery that led to printing presses, polio vaccines, and smartphones. . . . What would the deniers have us do? Ignore all the scientific discoveries that make our technologically driven world possible, things like the ability to rotate crops, pump water, generate electricity, and broadcast baseball?[8]

In summary, I echo the following thoughts from Casey Luskin, a past contributor for the website evolutionnews.org, about Bill Nye and others like him. Instead of steering all youth to science, Bill Nye just may be pushing a generation of geniuses right out of the field of science. Luskin says:

> Moreover, by adopting the patently false atheist-supremacist position that Darwin-skeptics can't do good science, Nye's rhetoric discourages many bright young Darwin-doubting students from entering scientific fields. In effect, Nye's own divisive prejudices and discriminatory attitudes toward Darwin-doubters may be hindering his own goals to inspire young people to become scientists and find scientific solutions to problems facing society.[9]

*So what exactly is the role of science?*

Science has two main purposes. One is to gain knowledge. There are a myriad of wonders to discover and learn about. For example, in studying the atom, one sees the nucleus with protons and neutrons. The proton has a positive charge and gives the atom its traits. The neutron has no charge at all. Thus, I had always thought it had no real importance. But it turns out that its job has major implications for the atom's very existence. What do like charges do? Repel! The nucleus would fly apart were it not for neutrons keeping the protons from repelling.

The more we study, the more we see how incredibly complex any item is. We now know that the atom's particles are made up of even smaller particles. Students are learning that bosons, leptons, and quarks are types of particles that make up protons, neutrons, and electrons. (For more information, research the standard model of the atom in particle physics.) I am

---

8. Quoted by Luskin, "Real Science."
9. Luskin, "Real Science."

sure you will agree that the atom looks very well designed and could not have just put itself together.

Sometimes the pronouncements of biological science that occasionally seem to trash creation and bolster evolution fall apart with further examination. There were headlines years ago about the discovery of "junk" DNA. Scientists found large patches of DNA that did not appear to code for anything. They therefore concluded that these were leftover remnants from our evolutionary past. Thankfully, someone studied them further anyway. These remnants have now been proven to not be junk at all; rather, they play a crucial role in the cell's operation, controlling the regulatory activity of other DNA parts. Therefore, sometimes science can actually be a friend to the creationist and foe to the evolutionist. So whenever evolutionists trumpet some new fact or fossil, stay tuned . . .

So while one role of science is to gain knowledge, another is to be helpful to mankind. The high standard of living enjoyed in many countries across the world is due in part to the role of science in creating the myriad technologies that convenience our lives. In this realm, science can help us help others. Drought-resistant crops, created by the science of genetics, help people of water-poor lands increase food production. Medical science heals diseases. Breakthroughs in energy production give us power with less pollution. This use of science is what one could call *ministerial*, to borrow a term from Martin Luther, who used it to explain the role of reason when studying God's word.[10] Science is the servant of mankind for the betterment of mankind.

That said, science is not the end-all used to judge the word of God. That would be a *magisterial* use.[11] People have used science to show supposed errors in the Bible, which are then used to "prove" that it is not God's word. This is not the role science is to take. But to understand how some do use science to try to prove God's word to be wrong, let's look at one Scripture passage commonly touted by the secular world.

Read Joshua 10:12–14. During a battle with the Amorites, Joshua apparently needs more time for a complete victory. He appeals to the Lord to make the sun and moon stand still, and "the sun stood still, and the moon stopped" (v. 13). A miracle indeed! But Bible doubters think they can stump a Christian with this passage. They claim that the Bible, which is said to be inerrant, does in fact contain an error here: everyone knows

10. Hein, "Reason."
11. Hein, "Reason."

## Does Neutral Ground Exist?

that the sun and moon do not actually move in the sky; rather it is the earth that rotates, making it only seem as though they are moving. Note that evolutionists use words like sunset and sunrise so therefore they are being just as unscientific. The truth is that the Bible uses poetic language, indicating in this case that God did indeed make the daylight last longer to aid Joshua to victory.

Although the Book of Jashar, a text of history or recorded events mentioned by Joshua in verse 13, is forever lost, one can find evidence from other ancient texts of this long day in Greece, Egypt, and other ancient nations. The American Indians, South Sea islanders, and others in the Western Hemisphere also tell legends of a long night.[12] So those living on the same side of world where the battle happened remarked the long day actually happened, while the ancients on the other side of the world told of a long night that occurred. Even more, consider that the ancient peoples did not read each other's legends or tales. Seems to be compelling evidence indeed.

*If two purposes of science are to help mankind and to gain knowledge, then how can science be so antagonistic to Christians who believe in Genesis as historical truth?*

It's important to understand that there are two types of science.[13] The first is *operational science*. This type gives us all our technology, medicine, and devices to extend our knowledge. It is what most of science actually is. The second type of science is called *origins science*. This type attempts to find out how we got here, how the earth got here, how the universe got here, and . . . you get the idea.

The website Answers in Genesis defines and explains origins science as "interpreting evidence from past events based on a presupposed philosophical point of view."[14] "The past is not directly observable, testable, repeatable, or falsifiable; so interpretations of past events present greater challenges than interpretations involving operational science. Neither creation nor evolution is directly observable, testable, repeatable, or falsifiable. Each is based on certain philosophical assumptions about how the earth began. Naturalistic evolution assumes that there was no God, and biblical creation assumes that there was a God who created everything in the

---

12. Morris, *Defender's Study Bible*.
13. Patterson, "Chapter 1."
14. Patterson, "Chapter 1."

universe. Starting from two opposite presuppositions and looking at the same evidence, the explanations of the history of the universe are very different. The argument is not over the evidence—the evidence is the same—it is over the way the evidence should be interpreted."[15]

Now we are getting close to answering why this problem of neutral ground is so very important. Here it is in a nutshell: there is no such thing as neutral ground. Jesus said much the same thing: "Whoever is not with Me is against Me" (Matthew 12:30). Does his statement leave room for any neutrality? Not at all. The reason why neutral ground cannot exist for the evolutionist is quite simple: "The mind that is set on the flesh is hostile to God" (Romans 8:7). The evolutionist's mind is set on the physical, while the Christian's mind is set on the spiritual found in Jesus. The Bible leaves no room for neutrality because neutrality does not exist.

It is very telling indeed when the evolutionist asks for open-mindedness and tolerance yet is viciously unreceptive and hostile to anything he or she doesn't agree with. The sole reason evolutionists ask for neutrality is so they can win the creation versus evolution argument. Evolutionists are never neutral, and they will ask you not to stand upon the Bible. But I encourage you to stand firm and to do so with the word of God as your foundation. So as stated earlier, the reality is that the evolution and creation debate is not a conflict between religion and science, but against two competing worldviews or philosophies.

This conflict is made more apparent if we let the evolutionists speak for themselves.

> We take the side of science *in spite* of the patent absurdity of some of its constructs, *in spite* of its failure to fulfill many of its extravagant promises of health and life, *in spite* of the tolerance of the scientific community for unsubstantiated just-so stories, because we have a prior commitment, a commitment to materialism. It is not that the methods and institutions of science somehow compel us to accept a material explanation of the phenomenal world, but, on the contrary, that we are forced by our *a priori* adherence to material causes to create an apparatus of investigation and a set of concepts that produce material explanations, no matter how counter-intuitive, no matter how mystifying to the uninitiated. Moreover, that materialism is an absolute, for we cannot allow a Divine Foot in the door.[16]

15. Patterson, "Chapter 1,"
16. Lewontin, "Billions and Billions."

## Does Neutral Ground Exist?

Read carefully what the author of the above quote, evolutionary biologist Richard Lewontin of Harvard University, is actually admitting. Some of what evolution proclaims is downright foolish. For example, take the evolutionary idea that sexual reproduction happened before single-celled life forms became multicellular life forms. This idea is taught in the current edition of the biology text used by many high schools across the United States. Students are taught that sperm and egg cells evolved before being produced by tissues of multicellular living things. This seems foolish to anyone who knows basic biology. But as Lewontin stated above, anything relating to God is ruled out. A good question to ask would be why this is so.

Note further that it is not the data or facts discovered by the methods and institutions of science that compel evolutionists to accept evolution as true. Rather, a prior commitment to materialism and material explanations is what makes the concept of life putting itself together seem valid. Why is there this steadfastness to material causes? Science cannot "allow a divine foot in the door," so it is hostile to anything that seems godly. These scientists are not true atheists, in one sense, as they worship at the altar of nature. Ask yourself just why the word *nature* is capitalized so often in science books and magazine articles. It is because nature is now taken to be a god that can do the miraculous all by itself, even though there is not one shred of evidence that it can do so.

Another telling quote comes from evolutionist D. M. S. Watson, who declared the following in an address to his fellow biologists at a Cape Town conference.

> Evolution itself is accepted by zoologists not because it has been observed to occur or . . . can be proved by logically coherent evidence to be true, but because the only alternative, special creation, is clearly incredible.[17]

However, to Christians, evolution claiming that we are simply mutated fish is "clearly incredible."

There is one more scientist to quote—Professor James M. Tour, one of the ten most-cited chemists in the world. His name is on thirty-six patents, and his most recent successes are in the areas of nanoelectronics, carbon nanovectors in medicine, and more. In other words, he is one practical and practicing scientist. Read what he said to a student who asked him about evolution at a talk he gave at Georgia Tech in 2012:

---

17. Quoted in Schlossberg, *Idols for Destruction*, 144–45.

## Is Evolution Compatible with Christianity?

> I don't understand evolution, and I will confess that to you ... Let me tell you what goes on in the back rooms of science—with National Academy members, with Nobel Prize winners. I have sat with them, and when I get them alone, not in public—because it's a scary thing, if you say what I just said—I say, "Do you understand all of this, where all of this came from, and how this happens?" *Every time* that I have sat with people who are synthetic chemists, who understand this, they go "Uh-uh. Nope." These people are just so far off, on how to believe this stuff came together.... Sometimes I will say, "Do you understand this?" And if they're afraid to say "Yes," they say nothing. They just stare at me, because they can't sincerely do it.[18]

Even more fascinating is that Professor Tour has convinced some evolution-believing scientists to change their minds on evolution. One person in particular whom his discussions on macroevolution influenced was Nobel Laureate Richard Smalley, who gave up his ideas on evolution shortly before his death. While Dr. Smalley accepted Old Earth creationism, which is the belief that the earth is billions of years old and that Genesis 1 and Genesis 2 are poetic accounts of the creation done by God, read what he said was the reason he stopped agreeing with evolution:

> Evolution has just been dealt its deathblow. After reading "Origins of Life," with my background in chemistry and physics, it is clear evolution could not have occurred. The new book, "Who Was Adam?" is the silver bullet that puts the evolutionary model to death.[19]

*Origins of Life* and *Who Was Adam?* were both written by Dr. Hugh Ross (an astrophysicist) and Dr. Fazale Rana (a biochemist). Did these two books alone change this scientist's mind? No, but they were surely influential, and that is one of the reasons I am writing this book. I hope I can encourage you to share what you learn with others, and perhaps, with the help of the Holy Spirit, more souls will be won for God's kingdom.

Various evolutionary ideas will be presented and swiftly dismantled in future chapters of this book. But for now, we must look at one serious consequence of a creationist agreeing to neutral ground when discussing science, morality, or even religion with an evolutionist.

---

18. Torley, "World-Famous Chemist."
19. Torley, "World-Famous Chemist."

## Does Neutral Ground Exist?

Let's look at James 4:4: "You adulterous people! Do you not know that friendship with the world is enmity with God? Therefore whoever wishes to be a friend of the world makes himself an enemy of God." All of our thoughts and beliefs, science included, need to be in line with the Bible. God takes this very seriously, and we ought to as well. Think now about this issue of neutral ground. Who is it that wants you to put aside the word of God, beginning with Genesis 1:1? It is none other than Satan himself. Being on the side of Satan has serious consequences indeed.

*Are you saying that anyone who teaches evolution is working for Satan?*

Evolutionists may not be working directly for Satan, but they are definitely not proclaiming the solid truth from God's word. In matters of operational science, any investigation or teaching of the wonders of God's creative genius automatically gives glory to him, whether or not the teacher acknowledges this truth. There are Christian teachers and professors who are strictly forbidden to teach publicly that God is indeed the Maker of all things, but outside of the school property, those teachers or professors may have a different tale to tell. However, if a teacher gives glory to godless forces of nature or evolution, this is an entirely different matter. There are those who specifically force students to deny that God was involved at all in any way.

Just as perspective or worldview determines how one interprets the data gleaned from studying creation, so one's worldview can be applied to any occupation or field of study. The question is, will you use that knowledge to further God's kingdom here on earth or will you hinder the spread of the Gospel? There is no neutral ground anywhere.

*Are there other places in Scripture where Jesus or God being the only side can be found?*

Colossians 2:3–4 says, "In [Christ] are hidden all the treasures of wisdom and knowledge. I say this in order that no one may delude you with plausible arguments." Here Paul is telling us that everything that relates to knowledge and wisdom finds Jesus at its core. It does not matter if one studies languages, literature, or economics. All knowledge or wisdom has Jesus at the most basic level.

Beware of those who speak of unbiblical ideas, such as macroevolution which is Evolution Wrong, or who say that there are many ways to heaven. Such lies come from Satan himself. It does not matter how convincing the argument sounds. What matters is whether the logic comes from God's word or from man. This is why you should continuously study the creation/evolution issue. Learn the arguments made by evolutionists so that you can identify their flawed thinking or the worldview presuppositions they use to support their ideas. It will indeed surprise them to deal with a prepared Christian who can show just why evolution in the macro sense (Evolution Wrong) is illogical as well as unbiblical.

*How can you avoid falling prey to the well-written explanations and reasoning of evolutionists?*

The solution is very simple. In Titus 1:9, Paul tells Titus that a pastor is to "hold firm to the trustworthy word as taught, so that he may be able to give instruction in sound doctrine and also to rebuke those who contradict it." This is excellent advice for any Christian to heed. Note that the verse says both to give instruction and to rebuke those contradicting it. Also remember that in Ephesians 6:17, the Bible is described as the sword of the Spirit, and in Hebrews 4:12, we are told that the word of God is sharper than any two-edged sword. One can defend and attack with a sword. I encourage you to learn to use the Bible to defend your faith, which does indeed teach that God is the Creator of all. Learn also how to use God's word to attack the false ideas and doctrines of the world.

# Chapter 3

# Definitions
# or
# What Does This Mean?

> "What is the meaning of it, Watson?" said Holmes solemnly
> as he laid down the paper.
> —Sherlock Holmes and Watson, "The Adventure
> of the Cardboard Box"

The word *definition* can mean "a statement of the meaning of a word, phrase, or term." If you already knew that, thank a teacher. Definitions are critical to the topic at hand. Two people can use the same word in the same sentence but mean entirely different things. This goes beyond the connotation and denotation concepts I once taught to my ninth grade English students. A quick refresher: denotations are the clinical dictionary definitions for words, and connotations are the feelings associated with those same words. For example, when one says, "That is so bad," one of two meanings could be communicated. Someone could be using *bad* as slang, indicating that the item referred to is very excellent or exciting. Another person could use the word *bad* to mean that the item is terrible or not desirable.

When dealing with scientific concepts, connotations of words usually do not apply. Yet, this does not mean that words cannot have more than one meaning in science; one must still be careful in defining one's terms so that clarity of ideas results. Let us tackle some of the troublesome words that cause misconstrued ideas between two people discussing the creation and evolution issue.

*What exactly does the word evolution mean?*

The former president of our state science education organization defined *evolution* as most textbooks do today. Shaking her head at me with a weary expression, she said, "Don't you understand that evolution is just change through time?" For a Christian (I will define this word also), there is nothing biblically or scientifically wrong with this definition. The problem is that what evolutionists say is not always what they mean. They mean something that is unbiblical and also unscientific in the traditional sense. Note that this definition of evolution (change through time) is so vacuous as to be meaningless. The evolutionist's change over time means that protocells eventually become people via mutations. The Christian's change over time means that the animals, which were aboard Noah's ark, came out after the flood and have diversified or adapted (evolved) into the forms we see today.

*How can a Christian respond to someone who believes in evolution rather than creation?*

Christians need to respond to evolutionists who use the phrase *change through time*, or *change over time*, by asking two clarifying questions: (1) How much change? (2) How much time? Once those questions are answered, the Christian needs to ask more specific questions. For instance, is evolution the finch beaks changing sizes during drought and rainy seasons on the Galapagos (as documented by Peter R. and B. Rosemary Grant, emeritus professors of ecology and evolutionary biology of Princeton University[1])? The evolutionist ought to respond affirmatively. Rains make the seeds that the birds eat softer, so their beaks shrink a bit because the seeds are easier to eat. Droughts bring drier, crunchier seeds; thus bigger and stronger beaks are needed.

Next, ask if evolution also means that atoms can become astronauts. Evolutionists will agree with this as well. Their definition of evolution has a high plasticity; it means just about anything they want it to mean. Evolutionists allow for this broad definition so that there will never even be a hint as to evolution not being true. Think about what might happen if evolution

1. Miller and Levine, *Biology*, 496.

## Definitions or What Does This Mean?

as an idea were found to be false. Some people, mainly evolutionists, would be out of a job. Then there is the problem of having to admit that one is wrong. But is it even possible that evolution could be wrong? Indeed, it is possible depending on what one means by evolution.

If the famous evolutionary biologist Richard Dawkins can claim that nature only looks designed, but it really is not, then the opposite is also possible. Nature really was created, and the great flood really did lay down fossils in ecological zonation order, but everything looks evolved only to a viewer who chooses to see it that way.

*Shouldn't we also define the word time, since so many evolutionists use that as part of their definition of evolution?*

*Time* is a word that is difficult to understand or define. One dictionary defines it this way: a non-spatial continuum in which events occur in apparently irreversible succession from the past through the present to the future. Time as a word is found in sayings. "Time is of the essence," so the adage goes. It is true in this case, as time is quite an essential element to the idea of evolution. Without billions of years, almost all evolutionists insist that evolution would be impossible. Christians who are biblical creationists follow the Bible's timescale, stating that the earth is anywhere from six thousand to ten thousand years old. But it is not just the Bible that speaks of a young earth. There is geologic evidence that is more properly interpreted as having a recent single catastrophic event origin, namely the biblical flood found in Genesis 6–9. Thus if the rock layers were laid down in one single cataclysmic event, evolutionists would have much less evidence interpreted as "old" and then there never were billions of years.

*What exactly is a Christian?*

The term *Christian* was coined within the first years after Jesus' death and resurrection. In Acts 11:26, we read, "And in Antioch the disciples were first called Christians." The term was first meant as derogatory; however, it obtained a more tolerable meaning as the years went on. But history aside, what is a Christian? Surprisingly, I am going to let an atheist define the term.

A reporter for *Portland Monthly Magazine* interviewed atheist Christopher Hitchens. During the session, the reporter revealed that she does not take the stories from the Bible literally; she believes the atoning work of Jesus' sacrifice on the cross is simply a story. Hitchens replied, "I would say that if you don't believe that Jesus of Nazareth was the Christ and Messiah, and that he rose again from the dead and by his sacrifice our sins are forgiven, you're really not in any meaningful sense a Christian."[2] Wow! I could not have said it better myself. It is true that there are many shades of Christianity. But I fully agree with many Christian teachers and apologists that as long as one holds the essentials of the historic Christian faith, then one may say that he or she is a Christian.

*Let's define science. It probably means more than bubbling test tubes and mathematical equations, doesn't it?*

This term needs quite a bit of space, as it really is a major concept. If you had to define *science*, what would you say? Better yet, what part of speech is it? Most people say that it is a noun. I have heard science educators state that science is a verb, fitting nicely with a hands-on approach to instruction. Some dictionaries use these phrases in their definitions: knowledge gained through experiment; a way of knowing something; or the observation, identification, description, experimental investigation, and theoretical explanation of phenomena.

In his book *Signature in the Cell*, Stephen Meyer points out that different fields of science utilize differing modes of operation. What is even more telling is that philosophers of science cannot agree on one single definition of what science is, due to the diversity of methods used by scientists. Thus, when an evolutionist tells you that intelligent design or creation is not scientific, you can ask him which definition of science he is using in making that judgment. Do all philosophers of science agree with him? Decidedly not.

Meyer continues to elaborate this "demarcation problem."[3] For example, some scientists use the typical lab experiment, but others do not. Some try to reconstruct events of the distant past while others confine themselves only to the present. Some try to discover laws governing nature while others simply try to understand how something works or what its parts are. Finally, some evaluate the explanatory power of others' ideas or statements. On the other hand, many creation scientists try to provide plausible explanations for how things work as well as how things came to be.

One can study proteins, DNA, or any other topic in science and not have to believe evolution is true whatsoever. Something's origin has

---

2. Hitchens, "Hitchens Transcript."
3. Meyer, *Signature in Cell*, 400.

nothing to do with its parts or what it does and how it does it. Testable and lab-related science is best when strictly confined to the present. Science tells us, "This is what it is, and this is what it does." Period! The further back we go when we attempt to scientifically examine the past, the less certain we are of anything, especially if no one today was alive when the event happened. This is exactly what evolutionists argue when debating the evolution/creation issue: no creationist was alive when the earth was formed, so how can we claim to know anything about it? We can agree with them that no human was there, but God was. He even kept a "lab notebook" of sorts and then dictated that notebook to Moses in the writing of Genesis. There will be much more to come about this book of God's in a future chapter.

*How do we define the beginning of things?*

Our final term to deal with is *origins*. How did this whole cosmos come to be? Here, the creationist is on solid ground. Genesis 1:1 says, "In the beginning, God created the heavens and the earth." But if creationists try to discover how it all began, we can use the same argument of the evolutionist—that no human was there to record events. Therefore, we are like all evolutionists when it comes to origins. All we have to go on is circumstantial evidence.

Whenever a scientist posits an explanation, it must not violate any experience or established scientific law. In regards to the origin of life, there is the law of biogenesis, which states that life comes only from life. Some evolutionists say that life originated from the backs of crystals or from a warm pond of molecules just doing their thing, which would be to bond with all the wrong things. As they do that random bonding they would not make life. But in this line of thinking, the evolutionist violates the known and well-established biological law: biogenesis, as we discussed above.

Evolutions will then cry that biblical creationists are now guilty of violating the law of biogenesis, but we respond with our belief that God has always been alive and will always be, and therefore life still came from life at creation. If the evolutionist then asks who made God, we can respond that according to philosophy, a First Cause needs no maker. Therefore, to ask who made God is an improper philosophical question. Furthermore, questions pertaining to God are outside the scientific realm. Besides, if God could be made, then whoever or whatever made God would have to be greater than God. And if God is not the First Cause of all things, then he is not God.

# Is Evolution Compatible with Christianity?

*How might an evolutionist respond to questions from a creationist?*

Sometimes a well-known evolutionist will make an interesting statement about creation. Read this quote from evolutionist Dr. Lynn Margulis: "The critics, including the creationist critics, are right about their criticism [that natural selection is not a mechanism for the evolution of new species]. It's just that they've got nothing to offer but intelligent design or 'God did it.' They have no alternatives that are scientific."[4] It is very possible that the questions of origins do not pertain to the realm of science, but to that of philosophy or religion; hence there will never be a "scientific" answer that can be experimentally or otherwise proven. Here is where Christians are advised to learn a bit about logic and using it to show that pure materialistic ideas fall short when dealing with origins. I recommend going to creationist websites to gain further understanding and knowledge with this.

*Where is a good place to start when looking for biological facts or concepts that cause problems for evolutionists?*

Evolutionists often dodge the topic of origins by saying that origins has nothing to do with evolution; evolution only has to do with life and its changes over time. They claim that the origin of life is a totally separate issue. But the creationist should reply with a resounding and emphatic, "No, it is not." One of the phrases I hope my students take away from my class is this: "If a process cannot start, it cannot continue." Obviously, if life cannot put itself together, then it would be impossible for cells to evolve from it. Besides, if the origin of life has nothing to do with evolution, then why did Dr. Ken Miller, an evolutionist and co-author of the biology text high schools have used for a number of years, include a chapter in his book on the origin of life in the major unit covering evolution?

In reality the issue of origins trips up many an evolutionist. When we study the cell, we find a host of biological machines doing actual work inside it. Each machine usually has parts similar to man-made machines in the real world . . . or should I say our big, non-cellular world. I suggest you begin with one of the more "simple" machines, the walking motor protein. As Figure 1 shows, it looks like something made to walk on two legs. If you Google it, you can even find video animations of this wonder in action.

---

4. Teresi, "Discover Interview: Lynn Margulis."

# Definitions or What Does This Mean?

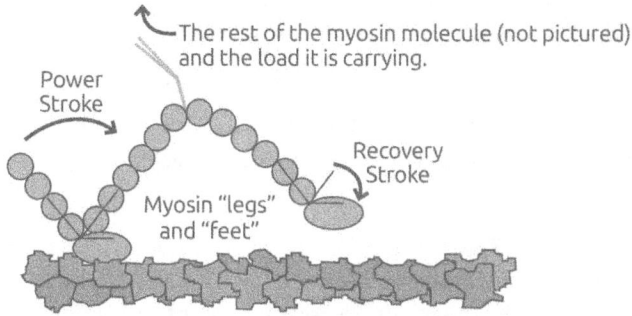

**Figure 1.**

My favorite is the one showing these marvels moving vacuoles (the storage sacks for a cell) along the cell's cytoskeleton or microtubules (the sidewalks and roadways inside a cell). The first animation of their existence and work was *The Inner Life of the Cell*, a video produced by Harvard University in 2006. When I first saw this animation, I recalled the now famous line of the 1931 movie *Frankenstein* when the mad doctor's creation came to life after the lightning storm. He cried out, "It's moving! It's alive!" Watch the video and see if you have the same excitement Dr. Frankenstein had.

These marvels of cellular engineering are quite astounding. Basically, the motor proteins are the delivery trucks moving various items back and forth inside each one of our cells. They move vital molecules along the microtubules inside the cell. If you have recently had biology, mentioning microtubules might bring to mind mitosis. During cell division, the chromosomes somehow move from the equator of the cell to the poles. Now you know what moves them. Powered by an energy source called adenosine triphosphate (ATP), the motor heads (feet) actually move as our legs do, one in front of the other.

*Just how important are these molecular motors to eukaryotic cells (the cells with a true nucleus, like those in our body)?*

Read what biophysicists at two universities in Munich have to say about it:

> Motorized transport proteins are one of the keys to the development of higher organisms. It is they that enable the cell to transport important substances directly and quickly to a specific location in

the cell. As bacteria cannot do this, they are not able to form larger cells or even large organisms with many cells.[5]

Note that the biophysicists claim these motor proteins are critical to developing (evolving) higher organisms.

Where in the world did these miniature machines come from? The next quote makes it plain.

> "Our results show that a molecular motor must take on a large number of functions over and above simple transport, if it wants to operate successfully in a cell," says Professor Matthias Rief from the Physics Department of the [Technical University of Munich]. It must be possible to switch the motor on and off, and it must be able to accept a load needed at a specific location and hand it over at the destination. "It is impressive how nature manages to combine all of these functions in one molecule," Rief says. "In this respect it is still far superior to all the efforts of modern nanotechnology and serves as a great example to us all."[6]

There you have it. Nature is responsible for this biological contraption.

The truth is the cell needs cellular machines of many types or it would die. A critical question needs to be asked. How did the first cells survive while waiting around for DNA mutations to build a machine that actually worked? Isn't a machine, or at least a designing intelligence, needed to build a machine?

Here again the evolutionists will attempt to explain how primitive machines were made with fewer parts from co-opted parts of other molecules doing the job at first because they cannot allow any hint of purpose or design. But this begs the question that if the primitive parts worked okay, then why bother evolving another more complex machine that does the exact same thing? Does it make sense to you? It makes no sense to me.

It is not that difficult to trip up an evolutionist with the issue of origins. Just ask simple questions. For example, I had a lot of fun at a science conference asking one evolutionist presenter with a doctorate degree to explain the origin of mitosis. I honestly forget now what his answer was, but I do know that his response was not even in the ballpark of the topic of my question. So after the session was done, I told him that he really did not answer my question, and he then referred me to the website talkorigins.org. Upon searching the website for an explanation on the origin of mitosis, I

---

5. Technische Universitaet, "Intracellular Express."
6. Technische Universitaet, "Intracellular Express."

## Definitions or What Does This Mean?

found only two short paragraphs. Basically it said that mitosis is thought to have started via "mechanical instability." Huh? The highly detailed process of mitosis began by the first cells simply coming apart? This is what counts as a scientifically sound idea worth testing? In chapter 11, we shall revisit the origin of life ideas presented by the evolutionist establishment and then refute each idea based on sound science.

# Chapter 4

# Christianity

## The Bible's or Dr. Kenneth Miller's?

*"It is a capital mistake to theorize before you have all the evidence. It biases the judgment."*

—Sherlock Holmes, "A Study in Scarlet"

One may rightly question how this Holmes quote fits with a chapter dealing with Christianity. It is my contention that many evolutionist scientists are not allowing the facts discovered to alter the theory or idea of evolution, especially the part of evolution dealing with origins. Now to be fair, science is always uncovering new facts and rightly so that theories are continually being modified or even abandoned in a few cases. It also is my contention that had Darwin and others knew about the complexity of cells and the impossibility for mutations to "design" anything, evolution as an idea for origins without God or the history of life on earth would have never gotten a foothold in the scientific domain. It is true that as new facts are discovered they are then shoehorned into the evolutionary paradigm instead of evolution being altered or even discarded in some cases.

This chapter deals with Christianity. I fully admit that I am a Christian who believes that the Bible is the inspired word of God. It is not man's word at all, even though men wrote it. This is what it claims itself to be as

1 Thessalonians 2:13 states. "And we also thank God constantly for this, that when you received the word of God, which you heard from us, you accepted it not as the word of men but as what it really is, the word of God, which is at work in you believers." This is helpful in joining the first sentence of the quote with the second. Am I biased then because I am a Christian and therefore interpret facts as pointing to creation? Yes, indeed. But then again, so are the materialistic/atheistic evolutionary scientists who claim that the facts show that evolution is true and that God had no part in the creation of the world or anything in it.

*But aren't there books that one can read to help figure out how Christians are supposed to integrate his/her faith with science?*

I strongly caution all Christians to be discerning when it comes to the books or websites they read. Although a title may sound helpful, or the author publicly proclaims to be Christian, the item itself may not correctly portray Christian doctrine. One such book is *Finding Darwin's God: A Scientist's Search for Common Ground between God and Evolution.* It's written by Dr. Kenneth Miller, who also happens to be the co-author of the biology text I used in my classroom. Dr. Miller is one of the premier biology text authors and has spoken at numerous science conferences dealing with the creation versus evolution issue. Chances are that if you have been educated in an American high school, you have read Miller's work in class.

Before continuing I must stress that I am not judging Dr. Miller in any way. Nor am I claiming to know what his motives are when saying things in his book. I will evaluate what he says and provide my thoughts on what is said by him. In Miller's book, the word *God* and all divine pronouns are capitalized. He also says that Christians need to "demand that non-believers carry the heavy logical burden of their absolute materialism all the way to its conclusion."[1] But Miller does not elaborate on what that conclusion is. Neither does he show us the logic of it nor how to use it against the materialist. Surprisingly, Miller elsewhere states, "When I have publicly advanced the idea that God is compatible with evolution, I find my agnostic and atheistic colleagues are generally comfortable with such ideas."[2] The ones I have spoken with are openly hostile to even the hint that God or a god could exist, even if he had nothing at all to do with this universe. It is

---

1. Miller, *Finding Darwin's God*, 271.
2. Miller, *Finding Darwin's God*, 220.

# Is Evolution Compatible with Christianity?

a mystery why Miller tiptoes around stating his personal beliefs and never specifically says what he believes.

*So what does Miller say that suggests his beliefs might not be biblical?*

The Bible clearly teaches that Christianity is the direct revelation of God to man in 2 Peter 1:21. "For no prophecy was ever produced by the will of man, but men spoke from God as they were carried along by the Holy Spirit." This one, along with the 1 Thessalonians verse confirms this. As the Bible is God's revelation of himself and his plan for mankind, it would make sense if Miller would use Scripture in a book he would write. After all even others claim that he is a Christian and considered by some to be an evangelical one at that.[3] But it is the absence of any Bible verses that gives me pause. Let us now examine some of his statements.

"I suggest that if God is real, we should be able to find Him somewhere else in the bright light of human knowledge, spiritual *and* scientific."[4] Find God somewhere else? Like where? Does he mean one can find God in other spiritual knowledge like Islam or Hinduism? And Miller, being a Christian, has a bizarre aversion to stating what Scriptures teach about the "bright light" of humanity. Worldly knowledge is said to be darkness. Read John 8:12, "Again Jesus spoke to them, saying, 'I am the light of the world. Whoever follows me will not walk in darkness, but will have the light of life.'" Or consider Psalm 82:4-5, "Rescue the weak and the needy; deliver them from the hand of the wicked. They have neither knowledge nor understanding, they walk about in darkness; all the foundations of the earth are shaken." These verses, as well as others, state that without Jesus, anything of the world is darkness. So how can Miller claim otherwise? Miller also says nothing about using God's word as a light. Yet Psalm 119:105 says, "Your word is a lamp to my feet and a light to my path." I can understand why Miller does not say these things in a biology text written for use in public schools, but why not in a book that he alone authored and thus had total control of?

Miller's most glaring denial of Christianity comes in the section of his book where he speaks about the grace of God.

> To a believer, grace is a gift from God that enables us to place our lives in their proper context—not by denying our biological heritage, but by using it in His service. To a believer, God's greatest gift

---

3. Forrest and Gross, *Creationism's Trojan Horse*, 53.
4. Miller, *Finding Darwin's God*, 267.

was to provide us with a means to understand, to master, and to do good using both the strengths and weaknesses of human nature.[5]

What is so misleading about what he has said one might ask. As you can see from the previous quote, Miller more than dropped the ball; he has forfeited the game. We can do good by using our strengths, yes. But he also claims that our weaknesses are to be used. But Scripture teaches that our weaknesses are sin. Read Hebrews 4:15, "For we do not have a high priest who is unable to sympathize with our weaknesses, but one who in every respect has been tempted as we are, yet without sin." This is saying that Jesus knows about our weaknesses, was tempted, and did not sin. We are weak, get tempted, and do sin. Can any sin improve the planet or help others? No. Also note Miller's claim that we are to use our biological heritage, since we have supposedly evolved from bacteria, to master and do good. How does knowing that one has evolved help us to do good? Certainly the Bible makes it crystal clear that on our own we are good for nothing. It is only by the Spirit dwelling in us that we are able to do good things.

Galatians 5:22–23 says, "But the fruit of the Spirit is love, joy, peace, patience, kindness, goodness, faithfulness, gentleness, self-control; against such things there is no law." No mention of our biological heritage there. But Miller's error concerning grace is more than just coming up with a new meaning for God's gift. He says nothing about God sending us his Son, Jesus, to be our atoning sacrifice for sin. *That* is God's greatest gift. Scripture defines grace so very easily. All Miller has to do is read Ephesians 1:7–8 to find that grace is found only in Jesus Christ: "In Him we have redemption through his blood, the forgiveness of our trespasses, according to the riches of His grace, which He lavished upon us, in all wisdom and insight." Titus 2:11 states it even more plainly: "For the grace of God has appeared, bringing salvation for all people." A simple idea does not appear, but a true God and true man does. Jesus was God's grace made flesh, and by his death and resurrection, we can be partakers of that grace if we believe by the Spirit's gift of faith.

As I read Miller's book *Finding Darwin's God*, I found no hint of Jesus anywhere. It seems that Miller places his smarts over God's. Has Miller forgotten or purposely ignored God's word, which says that "the fear of the Lord is the beginning of wisdom" (Ps 110:10), and that in Christ "are hidden all the treasures of wisdom and knowledge" (Col 2:2–3)?

---

5. Miller, *Finding Darwin's God*, 280.

# Is Evolution Compatible with Christianity?

*What are some other statements Miller makes in his book which are wrong about Christianity?*

Ken Miller not only gets God's grace wrong, but he also has some misconceptions of God that are unbiblical. Miller claims that Christian creationists disrespect God in this way:

> They hobble His genius by demanding that the material of His creation ought not to be capable of generating complexity. They demean the breadth of His vision by ridiculing the notion that the materials of this world could have evolved into beings with intelligence and self-awareness. And they compel Him to descend from heaven onto the factory floor by conscripting His labor into the design of each detail of each organism that graces the surface of our living planet.[6]

To me it hobbles his genius by reducing him to a being that cannot get what he wants from the start. It makes him the author of a woefully inefficient process (mutations) to eventually get intelligent beings who can have a relationship with him. Materials by themselves do nothing. If you have raw materials to make a car, and pile them into a heap, they will just sit there. Energy guided by intelligence is what makes or creates things in this world. God has limitless energy and intelligence far beyond what we possess. In my opinion it is more honoring to God when Christians simply accept God at his word and trust him to have told us the truth of how it happened.

*Do other Christians say things like Miller does?*

Of course they do. I chose Miller's ideas as he is one of the two author's of one of the most used biology textbooks, and secondly if one can deal effectively with Miller's biblically erroneous ideas, one can handle the other Christians espousing false notions. One such notion is that creationists limit God by making Genesis literal truth and saying that God did not use evolution to create. Others say that biblical creationists limit God by saying that the Genesis account is true. Just how does one limit a limitless God? I say it is better to acknowledge our own limitations by allowing God to tell us how he did it. But in the biology textbook, the authors limit God's power by saying he can only work through earthly materials over billions of years, leaving God with the only option of creating by waiting for the countless

---

6. Miller, *Finding Darwin's God*, 268.

mutations of genes. By saying this, they force God onto the factory floor to work in real time and overtime, constrained by the horribly inefficient and painfully cruel process of Evolution Wrong.

The creation of the world is mentioned more than one hundred times in the Scriptures, not just in Genesis. The specifics of other verses contradict the evolutionary fable foisted on us by the likes of Miller and other evolutionists, Christian or not. I say to you what God said to Job: "Brace yourself like a man. I will question you and you shall answer Me" (Job 38:3).

*What did God use to create?*

He used nothing—no chemistry, no physics, no anything. The concept is called creation *ex nihilo*. This means from nothing came the universe. Read what Hebrews 11:3b has to say: "So that what is seen was not made out of what was visible." Miller can quibble that atoms in gaseous states are essentially invisible, but the Scriptures are clear that there were no gasses from the Big Bang or first generation stars that made the heavier elements. Such an idea does not make sense in the light of God's truth anyway. The stars were not created until the fourth day, according to the Genesis account, so then how could they have been used by to God to "evolve" everything else?

*Okay, we have God as the Creator. That ought to be good enough. But does the Bible ever really say how He created the world?*

Did God use evolution? Let us examine the very word of God to see if He did. The first part of Hebrews 11:3 states that what is seen came about via the command of God. This is also stated in Psalm 33:6: "By the word of the Lord were the heavens made, and by the breath of His mouth all their host." Note that there is a difference between poetic imagery and poetic fact. The fact is that God has no mouth, but the fact also is that God used a command, and the universe existed instantaneously. Psalm 33:9 reinforces this: "For He spoke, and it came to be; He commanded, and it stood firm." Again, if a poem speaks of rain, this is poetic fact as well as scientific truth. There is no imagery used in verse 9. God used *nothing* to make everything. While Miller doesn't include this information in his book, Scripture is clear on this topic.

*Does* Finding Darwin's God *do any other damage to God's nature?*

Miller writes, God "establish[ed] a world that was truly independent of His whims,"[7] which implies that God has whims. Whims are sudden or capricious ideas or impulsive thoughts. God does not have whims. Whims belong to the human domain, and God is far above such nonsense. When God directs the course of history (his story!), he is not dictating but shaping events. Do we still have free will? Most assuredly we do. But does God know about these choices we make? Yes, he does. Time after time God intervened in the history of the Israelites. For example, Exodus 8:18–19 tells the story of the plague of gnats in Egypt. In Joshua 8:1, God tells Joshua that he had delivered the city of Ai into his hands. As you can see the classic cases of his interventions would be the exodus out of Egypt and the routing of the pagan nations in control of the Holy Land. These were not whims by any stretch of the imagination.

Miller also states that God has no idea what will happen in the future, that chance exists in the universe. For example, a coin flip is chance. According to what he writes, Miller believes God would have not known the outcome of the flip. Miller even ascribes his own biological conception to chance, which means that God had no idea that Ken Miller would be born in nine months. Yet, according to Miller, God has a plan for the Christian. But how can God have a plan for a person he has no idea about? Psalm 139:13 says, "For You formed my inward parts; You knitted me together in my mother's womb." No room for chance there. Isaiah states that the Lord formed him in his mother's womb in Isaiah 49:5. It is no error of biblical thinking to state that every person is formed in this way. God knows everything about us before we are born. We are *not* accidents of chance.

*So how does Miller merge evolution and creation?*

Miller eventually tries to explain how God and evolution worked together to make everything.

> Evolution does not move towards predictable outcomes. Given evolution's ability to adapt, to innovate, to test, and to experiment, sooner or later it would have given the Creator exactly what He was looking for—a creature who, like us, could know Him and love Him, could perceive the heavens and dream of the stars, a creature who would eventually discover the extraordinary process of evolution that filled His earth with so much life.[8]

---

7. Miller, *Finding Darwin's God*, 269.
8. Miller, *Finding Darwin's God*, 238–39.

I have already discussed the word *nature* capitalized in the middle of a sentence. The above quote from Miller gives us a clue as to why would an author would do this. Evolutionists must deify nature. To them, Nature is god. Look at the intelligence of nature using Miller's own words. Miller says that nature can test, innovate, experiment, etc. Nature was the intelligent designer. Perhaps what bugs evolutionists so much is the truth that only distinct beings with capable intellects are intelligent designers, who can test, innovate, experiment, and so on. Maybe that's what God meant when he said, "Let us make man in our image" (Gen 1:26). It is an interesting idea to ponder that we are intelligent designers as we were created to be like the intelligent designer, who is God.

Consider the glaring contradiction or oxymoron of Miller's quotes above. On one hand, evolution does not move toward a predictable outcome, nor does it have a goal in mind. On the other, Miller says that this process with no goal in mind would eventually give the Creator what he wanted. Isn't something you want a goal? That the Creator can have a goal but the process used by that Creator is not goal-driven makes no sense.

In his book *Finding Darwin's God*, Miller comments on ape behavior data that seems to have showed that a good God cannot exist. Miller disagrees with the conclusion by saying that he reads the data another way. Thanks for the loophole, Dr. Miller. Now biblical and intelligent design scientists can disagree with evolution by simply reading Miller's and others' data "another way."

*Are there others who merge God and evolution?*

Like Miller's *Finding Darwin's God*, which states that God used evolution to do his work, other books have espoused this notion as well, including a book my wife inadvertently purchased for my growing creationist library. Its title was intriguing—*God Did It, But How?* by Robert B. Fischer, who has a PhD in analytical chemistry.

Fischer has this to say in the preface: "Much of what is written and taught within evangelical Christian circles takes one particular viewpoint, while what is written and taught in some liberal circles (and also assumed in virtually all other non-evangelical circles) takes another viewpoint." So far, so good for Fischer. He continues: "The former is so unacceptable scientifically and the latter so unacceptable biblically that the controversy turns into emotional outbursts and senseless ridicule emanating from both sides."[9] And that's where his book gets interesting.

It appears that Fischer wishes to have compromise between the two sides. It has been said that a compromise is a decision that makes no one happy. Evolutionists, some of whom are abject materialists, will never allow a divine foot in the door, as Lewontin said.[10] Christians, on the other hand, are to hold firmly to God's word as it is written. There is no compromise possible. Show me where Jesus compromised with the Pharisees. Tell me where God compromised with Pharaoh. In the book of Revelation, will God compromise with Satan? The answer is no. Back to Fischer's quote, he is half right. The emotional outbursts are true, and evolution is unacceptable biblically. But I take issue with the statement that creation is scientifically unacceptable.

We know that messages come from intelligent minds, not from matter. Messages contain information and therefore information only comes from a mind and not matter. These first two statements can be scientifically tested and are no-brainers in real life experience. As DNA contains information and instructions (or messages) for protein synthesis, it is scientifically sound to say that the origin of DNA and of life came from an intelligent designer. I hold that the designer is God, as it says in Genesis. As will be stated later on, the miracles of creation do not violate known law, but operate at a higher law.

*Is there another section where Fischer limits God, and what is the danger of this?*

Fischer, along with all other theistic evolutionists, feels that God at times used both the supernatural as well as the natural to accomplish his means. True enough, for when the exodus happened, the Bible mentions a "strong east wind" (Exod 14:21) that drove back the waters of the Red Sea. This is God using nature, so the exodus is not such a big deal—say the compromisers. Wait a moment! Fischer and others ought to consider the details also mentioned in the Bible. Read on to Exodus 14:22: "The people of Israel went into the midst of the sea on dry ground, the waters being a wall to them on their right hand and on their left." Sure God used wind, but not just any wind. He had the wind make walls of water, leaving the ground

---

9. Fisher, *God Did It*, 1.
10. Lewontin, "Billions and Billion."

## CHRISTIANITY

of the sea dry. Those details point strongly to a very unnatural and fully supernatural event.

Theistic evolutionists not only play fast and loose with Bible text to make it read the way they wish, they also have problems of a scientific and theological nature as well. First of all, the science they so dearly love (the science they say rules over the Bible—because the Bible must change, not science) tells them that people do not rise from the dead. What about the resurrection of Jesus? Liberal theologians take care of that by saying Jesus only spiritually rose from the dead. If that is true, then why did the disciples find the tomb empty? Why did Jesus tell Thomas to touch his hands and side? You cannot touch a spirit. Jesus' body rose from the dead, just as all people will do one day. Believers' bodies will enter heaven and those who do not have Jesus as Savior will be cast into eternal punishment.

Another problem liberal theologians have is that, like Miller, they hobble God. They say he needs some help, as he can't do it alone. He needs nature to help him out. During a graduate school lecture, Miller recalls in his book a give and take between a presenter and a student. The student accused the presenter (who was from a religious university) of impropriety because he mentioned the evolutionary implications for a protein. Miller quotes the presenter's response: "If you deny evolution, then the sort of God you have in mind is a bit like a pool player who can sink 15 balls in a row, but only by taking 15 separate shots. My God plays the game differently. He walks up to the table, takes just one shot, and sinks all the balls. I ask you which player, which God is more worthy of praise and worship."[11] Miller says that he smiled at the response.

I would not have smiled at all. I would have risen my hand and said, "Neither one is worthy of praise and worship. Once in a while even a pro player can sink all the balls on the first shot." The biblical God simply speaks a table with racked balls into existence and speaks the balls to move to the pockets in numerical order around the table pockets. But this is merely a parlor magician's trick. God does not stoop to such feats. God spoke, and the universe simply came to be instantaneously. God loved each one of us enough to give us a choice to love him back. God loved everyone enough not to leave anyone lost in his sins. But to satisfy his perfect justice, he sent his only Son into the flesh to take each person's death sentence so that all may spend eternity with him. *That* is the God who is "worthy of praise and worship."

11. Miller, *Finding Darwin's God*, 284.

Another issue compromisers tend to sidestep is that of sin. What really is sin? If we are the product of evolution, whether by nature or by God's design and control, then sin is a non-issue and there really is no true right or wrong. Or sin is God's fault because he caused it to happen. The truth is that sin is our fault alone. Scripture says this in 1 John 1:5, "God is light, and in Him there is no darkness [sin] at all." What is really fun is to have an evolutionist explain why murder is wrong. Ask why, and he will respond by saying it is against the law. Answer back that if we change the law, then murder should be okay. He'll probably shoot back that that change will never happen. Tell him about Germany in 1939–1945. Remind him of Stalin or Mao. Yes, Christians have done their share of killing, but the atheists have way more blood on their hands as in millions more murders than we Christians do.

Finally, one major problem the compromise side has is a fairly serious one. Recall that God says that everything was good after each day of Genesis 1. At the end of the seventh day, it was very good. Had evolution been in operation for millions of years during "creation," then there would be animals dying, falling ill to disease, killing one another, and so on. Does this sound like paradise to you? And what about in the book of Revelation where God restores everything to the way it once was—except what was then is what we have now. So what did God mean by restoration? If the millions of years of evolution is somehow like heaven, I choose to go to the heaven described in the Bible where the lion lies down with the lamb and God wipes away every tear from my eyes!

*What is some biblical evidence that the Bible is indeed the word of God and does other evidence besides using the Bible exist?*

The Bible is what it says it is, namely the very words of God Almighty, the only one true God, Father, Son, and Holy Spirit. But where in Scripture does the Bible teach that it is God's word? Read what Paul commended the Thessalonians for in 1 Thessalonians 2:13: "When you received the word of God, which you heard from us, you accepted it not as the word of men but as what it really is, the word of God, which is at work in you believers."

Next, one needs to know the reason for the Bible. Why would God bother to have anyone write down his words? There really is only one reason: Jesus. He said so himself. After his resurrection, Jesus said, "Everything written about Me in the Law of Moses and the Prophets and the Psalms must be fulfilled" (Luke 24:44b). Earlier in his ministry, he was transfigured to have three disciples get a glimpse of his true nature and glory. Who appeared with him? Moses and Elijah did, representing the Law and the Prophets respectively. But again, why those two? Since they wrote about him, they were discussing his departure from this earth (death and resurrection). The Scriptures are written so that you, reader, may believe that Jesus is the Messiah, God's Christ, and believing in his name, have eternal life. John wrote about this very idea in John 20:30–31: "Jesus did many other signs in the presence of the disciples, which are not written in this book; but these are written to that you may believe that Jesus is the Christ, the Son of God, and that by believing you may have life in His name."

Concerning other sources dealing with the Bible, you would do well to read books by Josh MacDowell, Lee Strobel, or others like them. These authors used to be Jesus haters and tried to prove the Bible wrong. The more they tried, the more they began to believe. Go to Youtube and view debates between Christians such as John Lennox or William Craig and atheist opponents. Learn from the atheists' questions and attacks and the Christians' responses. Ravi Zakarias has some excellent clips as well.

Time now to wrap it up. The following are the reader's three choices: (1) Have man's wisdom and methods reign supreme. Science trumps the Bible every time. It is not God's word by any stretch of the imagination. (2) Have God's word mean something, but interpret it in the light of man's knowledge and oh-so-perfect wisdom (sarcasm intended). This way, you can make the Bible mean just about anything you want. (3) Have God's word as it is. Read it as the style of literature demands. If it is poetic, and most of Genesis is not poetry as the verb forms are mostly narrative, then you will find truth stated in imagery form, but it is still truth. If it is historical narrative, then you know it to be actual true events.

This, then, is the bottom line. You basically can believe the evolutionist who did not witness creation, or believe the God who was there and cannot lie about what happened. Logically how we got here is a matter of faith, but the choice to believe that humans know the actual truth requires a great deal more faith than the choice to believe in God revealing that truth.

# Chapter 5

# Is Theistic Evolution a Valid Compromise?

> "The more outré and grotesque an incident is the more carefully it deserves to be examined."
>
> —Sherlock Holmes, "The Hound of the Baskervilles"

To me, the notion that God used evolution to do his creative work is indeed an outré, an outlandish idea, and quite an odd one that cries out to be examined. As mentioned earlier, the idea that God had to use anything to get his work done makes him impotent—no longer *the* God but *a* god, if that. To join him, the supernatural, with a natural process is an oxymoron indeed. As said by the secular biologists, evolution is quite able to do the miracles of putting atoms into molecules, molecules into cells, cells into simple life, and so on until humans come to be. In other words, God is unnecessary. The late William Provine, professor of biology at Cornell, said it best in the movie *Expelled*.

> It starts by giving up an active Deity, then it gives up the hope that there's any life after death. When you give those two up, the rest of it follows fairly easily. You give up the hope that there's an immanent morality. And, finally, there's no human free will. If you believe in evolution, you can't hope for there being any free will.

## Is Theistic Evolution a Valid Compromise?

There's no hope whatsoever of there being any deep meaning in life: We live, we die, we're absolutely gone when we die.[1]

*Some Christians believe that God used evolution to do his creating. What's the problem with that?*

There are many problems with this, as hinted at earlier. One way to begin unpacking the notion of theistic evolution is to look at what self-described evolutionist Christians have to do to maintain this duality of thinking—that God is God and yet he still used evolution. In *Shadow of Oz: Theistic Evolution and the Absent God*, Wayne Rossiter says there are three ways evolutionist Christians can do this. First, they could wrap Christianity's tenets tightly around a purely evolutionary center. They also might erect artificial barriers between religious ideas and scientific beliefs. This makes sure that they never have to think about obvious contradictions even when they write them in print. The final way is to "push God into the distant and undetectable cosmic background so the universe only looks random (but it isn't)."[2] As you read the rest of this chapter, the reasons why Christians ought not do this type of thinking will become clear.

*What is an example of theistic evolutionist contradictory ideas?*

Some of the best examples come from Kenneth Miller, who we examined in chapter 4. One example is when he ridiculed Rick Santorum's statement that evolution tells us we are one of nature's mistakes, writing that it made him [Miller] "shake [his] head in amazement that someone living in the midst of the molecular revolution in biology could dismiss the creative power of evolution as nothing more than a series of mistakes."[3] Too bad Miller doesn't read the biology textbook he co-authored. In this same book he or his partner writes about man's genome being "riddled with useless information, mistakes, and broken genes."[4]

Another quote of Miller's or Levine's comes across as being quite atheistic: "Of course there has never been any kind of plan to evolution, because evolution works without either plan or purpose . . . Evolution is random and undirected."[5] Now match this with what Miller writes in one of his

---

1. *Expelled*.
2. Rossiter, *Shadow of Oz*, 9.
3. Miller, *Only a Theory*, 153.
4. Miller, *Only a Theory*, 96–97.
5. Miller and Levine, *Biology*, 658.

several books on the topic of evolution and faith. "People of faith . . . know that ours is a universe willed by God, and that our presence within it is part of his plan and purpose."[6] It sure sounds like Miller is trying to have it both ways. Either that or he is trying to fool someone.

*So how did this meshing of evolution and Christianity start?*

Some say that there were hints of theistic evolution as early as Augustine (354–430) or Aquinas (1225–1274). But if one reads what Augustine actually said, the truth would be obvious: "For as it is not yet six thousand years since the first man, who is called Adam, are not those to be ridiculed rather than refuted who try to persuade us of anything regarding a space of time so different from, and contrary to, the ascertained truth?"[7] Aquinas was not in the theistic evolutionist's camp at all. Consider his quote: "Nothing entirely new was afterwards made by God, but all things subsequently made had in a sense been made before in the work of the six days."[8]

Modern theistic evolutionists have ideas that are not much different from their founders, Pierre Teilhard de Chardin and Howard J. Van Till. Teilhard came up with the term "creative transformation." This means that God not only began creating, but also still continues it to this very day. That there is an ever increasing series of developments or transformations, as Teilhard called them, for the universe and the items in it, even to the point of saying that one of them was when mankind got his soul via evolution. To his credit, Teilhard did have a few events, such as the origin of the universe or the appearance of the first life, as being direct creative actions by God. But, generally speaking, anything that "evolved" like the first amphibian from the fishes would be classified as a creative act, as God would work through evolution to do the creating.

Van Till is like-minded for the most part, yet holds some other views, such as the death of men and animals before the fall. He says this:

> It is an incontrovertible scientific fact that there is a long history of life and death for periods of billions of years before people like you and I appeared on earth. So physical death before the fall must be accepted as a fact of science.[9]

---

6. Miller, *Only a Theory*, 154–55.
7. Augustine, *The City of God* 18.40.1.
8. Aquinas, "On Things," 3.
9. Stambaugh, "Death Before Sin?"

## Is Theistic Evolution a Valid Compromise?

Consider what this means in light of Genesis 2:17 when God tells Adam and Eve that the day they eat off the tree of the knowledge of good and evil, they shall die. So what? Big deal. According to some, humans had been dying for millions of years prior to that event in the garden of Eden. Also note that science has no notion of "the fall," so Van Till's statement does not make sense.

Van Till believes that God basically started the universe with its own creative potential. So God sort of wound it up and let it go on its own. Van Till calls this idea the universe's formational economy.

> The formational economy of the universe is sufficiently robust to make it possible—without need for occasional episodes of form-imposing intervention by any extra-natural agent—the actualization of every category of physical structure and biologic organism that has ever appeared in the universe's formational history.[10]

Like other theistic evolutionists, he is basically calling Genesis 1–3 a fairy tale at best or calling God a liar at worst.

### *What is so bad about theistic evolution?*

For one, it makes God the author of all sorts of suffering and misery. Please note that being the author or instigator of suffering is far different than allowing it to occur. For example, your parents allowed you to fall when you learned how to walk or ride a bike. Did they make you fall or cause it to happen? No. They allowed it to occur. As evolution can only move to where we are today via major changes in DNA, and as those major changes are only caused by mutations, and if we look at today's children born with genetic issues that no one would call a plus for the species, theistic evolutionists can conclude that God is then responsible for causing those children to suffer through life with disabilities. Further, it is true that most mutations are not even close to being adaptive to the environment; therefore, they would also say God is using a very inefficient process that really does not work at all.

Second, theistic evolution is not at all compatible with science to say that God is behind it all or that God used a process that he knew would produce something like us eventually. Evolution must be a totally random process that is in no way goal-directed or forward-looking. True evolutionists hold that man is just an "oops!" in the natural world, that there is nothing special about us, that someday we, too, just might go extinct. As I often

---

10. Van Till, "Is Creation?," 232.

said to biology class, "If evolution in the broad sense is true, then we are nothing more than mutated pond scum."

It would seem that the so-called Christian evolutionists want to have their cake and eat it too. Keith Stanovich stated so much when he wrote,

> The mistake that moderate religious believers in evolution make (as do many people holding nonreligious worldviews as well) is that they assume that science is only going to take half a loaf—leaving all our transcendental values untouched. Universal Darwinism, however, will not stop at half a loaf—a fact that religious fundamentalists sense better than moderates. Darwinism is indeed the universal acid—notions of natural selection as an algorithmic process will dissolve every concept of purpose, meaning, and human significance if not trumped by other concepts of equal potency . . . If you believe in a traditional concept of soul, you should know that there is little doubt that a fuller appreciation of the implications of evolutionary theory and the advances in the cognitive neurosciences is going to destroy that concept.[11]

He is basically saying that there is no 50/50 split or give and take when dealing with materialistic science and theological ideas, especially in the realm of creation/evolution. At least the atheist evolutionists are honest in that regard and say that there is no middle ground of compromise. Darwinism, as well as any modern ideas of macro-evolution, must never give an inch to notions of the soul or God. Please know that the full-blown atheist evolutionists do not agree at all with the theistic ones.

*Do theistic evolutionists also deny other tenets of classic Christianity?*

It is dangerous to generalize, but let this next example prove a point. Karl Giberson, while commenting on Norman Geisler's testimony during the public education battle over creation and evolution in Arkansas, said,

> Unfortunately [Geisler] embarrassed himself in Arkansas, becoming the brunt of countless jokes in the media . . . Geisler, like most fundamentalists, believes in a literal devil, the biblical Satan, and in demons. In this deposition for the trial he stated that he had "known personally at least twelve persons who were clearly possessed by the devil."[12]

---

11. Stanovich, *Robot's Rebellion*, 8.
12. Giberson, *Saving Darwin*, 101.

## Is Theistic Evolution a Valid Compromise?

Do Giberson and others like him not even realize that in mocking beliefs such as these he is mocking Jesus as being a "fundamentalist" since Jesus not only believed in Satan and demons, but routinely cast them out of people during his earthly ministry? So, yes, theistic evolutionists do deny the very things that Jesus himself taught and believed.

Finally, there is the concept of why judgment day is coming. To the true, Bible-believing Christian, the reason Jesus is coming back is to: (1) Take those who believe in him to eternal glory. (2) Judge the wicked unbelievers and cast them with Satan and all his angels into everlasting flames of hell. (3) Restore creation to what it once was by making all things new. Rossiter correctly points out,

> Theistic evolution puts the God-man project on its head, holding that creation emerges from chaos towards perfection, rather than it being in a state of continual decay. As we're about to see, this also requires that the theistic evolutionist abandon the existence of Adam and Eve, as well as the fall. But if we abandon the corruption of creation, then it follows that redemption is unnecessary . . . If this world is heading for perfection, then God does not need to make a new one.[13]

In summary of the above, the theistic evolutionist does not really need Jesus in any way shape or form. For if God used evolution, then there was never a real Adam and Eve. That means to call Jesus the "Second Adam" is ludicrous. As the world and humanity are "evolving" towards perfection, then Jesus' death and resurrection are not at all needed. It is also crucial to understand that the ultimate insult to God is that they are calling him a liar as he was not truthful about the beginning, and also he is lying about the end.

*Does Rossiter deal with any other flaws of evolution that theistic evolutionists continue to use?*

Rossiter does a fine job of demonstrating circular reasoning, a favorite tool evolutionists use. Many times the evolutionists will use this tool especially in high school biology books, as most students are not trained enough yet to see the logical fallacy of this tactic. Two facts or evidences that are supposed to support evolution are the fossil record and the tremendous diversity of life. When one asks for the reason that explains the existence of these facts,

---

13. Rossiter, *Shadow of Oz*, 70.

the answer is evolution. When one then asks how they know evolution did these, the answer is to look at the fossil record and the great diversity of life.

Please remember that the data one is trying to explain cannot be part of the explanation. In both of the above cases, the former requires the latter to be true before the former occurred. This then creates a circular argument that is neither logical nor scientific.

Rossiter also trashes the idea that one can diagram proof of evolution using the -ogram. If you are a biology student, examine your textbook to see if it has a cladogram, dendogram, or a phlogram. These are charts or trees that are said to support the evolution paradigm. If it has been a while since you were a biology student, these diagrams try to show the relationships to animals based on one common trait. Go to Answers in Genesis website and search "cladogram" to find articles about it. Consider that the evolutionists design a tree-building algorithm and then put data into it. The program produces a tree—no surprise really. It really cannot do anything else but that. What is so ironic is that they have used an intelligently designed program to show that there was no intelligent design involved with nature.

There is a final item that one needs to keep in mind. Evolutionists are human and as such will do anything to get the data to say what they want. They will even go so far as to remove any data that produces a conflicting story. Rossiter points out that in 2006 Rokas and Carroll, of the University of Wisconsin-Madison, chose to delete 35 percent of the single genes from a study as "those genes produced phylogenies at odds with conventional wisdom."[14] It just might be that conventional wisdom, as far as Evolution Wrong is concerned, is not that wise if you have to cheat to do it.

We close this chapter with a startling revelation. Most modern biologists are saying we have entered a post-Darwin era as far as evolution is concerned. What shocked me was that most of what Miller and Levine hold as scientific fact is now considered passé. Read this quote of a recent critical review of present-day evolutionary theory: "Many biologists feel that the foundations of the evolutionary paradigm that was constructed during the 1930s and 1940s (Mayr, 1982) and has dominated Western views of evolution for the last 60 years are crumbling, and that the construction of a new evolutionary paradigm is underway."[15]

14. Rokas and Carroll, "Bushes," e352.
15. Jablonka and Lamb, "Soft Inheritance," 389.

## Is Theistic Evolution a Valid Compromise?

Another quote I found interesting was from Melanie Mitchell, an expert in systems complexity. "The questions of how, why, and even *if* evolution creates complexity . . . are still very much open."[16] It really is too bad that science is still trying to find a natural explanation for what is in all probability due to its improbability, an act of a supreme intellect who the Christian calls *Abba* Father. One thing is certain, and that is most of what we have been taught in our high school or even our college classes in biology is flat out wrong when it comes to the creation of our universe and everything in it.

---

16. Mitchell, *Complexity*, 273.

# Chapter 6

# Philosophy 101

> "But is not all life pathetic and futile? Is not his story a microcosm of the whole? We reach. We grasp. And what is left in our hand at the end? A shadow."
> —Sherlock Holmes, "The Adventure of the Retired Colourman"

Before we discuss the nature of science, it is important to understand that the chain of dominoes goes like this: religion influences philosophy, which influences or is a worldview. Worldview is also known as metaphysics. Most think worldview is too esoteric for students and others to think about or comprehend. One reason for this is that most philosophers of science are deep thinkers. Deep thinkers sometimes don't have the skills to communicate to the average person. This means that they have done a poor job at times of explaining it to the everyday public. As Dr. Michael Ruse said on a *Firing Line* debate years ago, "I've been told I've got a minute and a half to make an opening statement. I'm a philosopher. I've never said anything in a minute and a half."[1] One could also read one of St. Paul's famously long sentences from one of his letters. Take a look at Ephesians 1:3–14. In most versions, it is broken down into a number of sentences. The truth is in the Greek it is one single sentence. This does not mean Paul is hard to understand at all. What it does mean is that he is very thorough in his discussion of a topic. I would say that St. Paul is one of the few deep thinkers that can

---

1. "The Firing Line."

actually be understood by most anyone, provided he reads one short phrase at a time and then puts the ideas together in his head.

Philosophy is certainly challenging, but like everything else, it has some fairly basic premises that can be easily explained and understood. Philosophy, or worldview, is simply why you are the way you are. Another way to say this is that it is why you see things the way you do. This worldview is created by one's religion, upbringing, and experiences.

*How did the Christian evolutionist become so very different from a Christian creationist?*

The Christian evolutionist believes that the Bible is just a bunch of stories cobbled from other Middle East pagan religions. For example, most world religions have a story about a great flood—some religions even have a large boat involved. He further holds that the Bible is simply a nice book of moral fables and has nothing to do with our day-to-day lives. After all, as he has been told, it is so very out of date. Plus, the writers of the time were so ignorant of scientific truths we know today. Additionally, the Christian evolutionist might have been raised with no biblical instruction, as his parents wanted him to choose his own religion or have none at all. Thus they would not be guilty of indoctrinating their child. Of course, the Christian evolutionist does not believe in a literal hell, as a loving God would never send anyone to hell. Finally, there are that person's experiences that provide the shaping of that view. He encounters Christians that hold to historic truths, but have never been taught apologetics or defending one's faith. He asks questions and makes statements for which nominal Christians have no response. He is now convinced that he is right, and it will take an act of the Holy Spirit to help him remove the blinders so tightly attached to his spiritual eyes.

*How can we tell who is wearing the blinders?*

At this point, I would have to add that the other side contends that it is people like me who have the blinders on. One question I would ask is this, "Does the one who is correct have the blinders on or does the one who stubbornly does not accept the truth have the blinders?" Jesus said, "I am the way and the truth and the life" (John 14:6). He also said in John 17:17, "Sanctify them by the truth; Your word is truth." There you have it. If one states what is in the Bible, then one is speaking the truth and it is the other who is wearing blinders.

# Is Evolution Compatible with Christianity?

I was raised to believe the biblical truths of God's word. In college, I was finally challenged to believe *and* know why I believe it. Remember the story I told at the beginning of this book? My professor knew most of us claimed Christianity, but in class he asked us to explain why we were Christian. Most of us said that we were raised that way. He then said, "So what makes you think that just because Mom and Dad are Christian that you are one? Care to show me anywhere in Scripture where 'coat-tail Christianity' saves you?" Needless to say, we all began digging through the Scriptures to back up our thoughts. Not surprisingly, many of us found that a few of our other thoughts were not biblical as well.

During his class, I began to try out other religions by dipping my toes into their waters, keeping my other foot firmly planted on the rock. I discovered that Christianity is unique for several reasons. Our Scriptures are written by over fifty different persons, yet the theme stays the same. The reason that this is possible is that there is only one real author of the Bible, and that is God. Another point is that we are the only religion where our God does everything for us, and we cannot do anything by ourselves except reject him. We are the only one to have God die in our place. Finally, we are the only one with an empty tomb. Now you can hopefully see that any other religion is man's attempt to reach God. Christianity is God's way to save humanity from hell.

## *What does all this have to do with science?*

Good question. Science is more than just obtaining facts about the world. Facts alone are only pointers in a certain direction. What a scientist must do is to interpret the facts. The interpretation is what the facts are telling him. What scientists face when examining nature is similar to a Sherlock Holmes story. Both Watson and Holmes saw the same facts. Both had individual interpretations, but Holmes got it right. Likewise, two different scientists can look at the same facts, but hold completely different interpretations. Why is that? Is one like Holmes and has more knowledge? Is it possible that one is right and one is wrong?

The answer hinges on the worldview of each. One holds that evolution, both Evolution Right and Evolution Wrong are true. In this case, evolution is a paradigm or framework on which to hang one's facts. The other holds that the Bible is true and interprets any facts though that lens or by that light. (Remember Psalm 119:105: "Your word is a lamp to my feet and a light to my path.") For those who think that evolutionists don't do such things, recall the words of the late evolutionist Stephen J. Gould, who stated that facts were always understood via theory's light.[2] Actually, to be any practical good, science needs to be correct and searching for truth. If such is the case, then theory ought to be adjusted by the light of facts. Sadly, the evolutionist never changes evolution. It remains unchanged and the facts are usually explained away by some fantastically illogical process, like mixing to explain how fossils of the wrong ages get into sediment layers absent of any other evidence save for the dates being "incorrect."

*Isn't it true that there is no such thing as absolute truth in science?*

If so, then no matter what facts are dug up, found in the flask, or seen on a DNA chromatography paper, one ought to ask: why is it that those facts always are said to confirm evolution? True, the same could be said of a creation scientist doing his work in that his facts will always support creation or at least intelligent design. But these two are merely interpretations. Again, is there any practical value for evolution? Did it put man on the moon? Did it give us any of our technologies or lifestyle? Does it really matter to the planet and our nation going green? Of course not. Remember, Christians, that the same can be said of creation, with the exception of going green. Back in Genesis, God told Adam that he was to be the caretaker of the garden. It sure seems to me that God would want Christians to be equally responsible for taking care of the world or environment now. Just understand that most environmentalists soundly reject creation and thus, ironically, will not accept that they are fulfilling God's command to humanity.

*Why is there such uproar over creation and evolution?*

It is true that this fuss never occurs when one studies mitosis or photosynthesis. The heated discussions in classrooms do not happen over the ideas of Mendel, Pasteur, or any other non-evolutionary biological scientist. The whole issue comes down to a simple question that Jesus asked his disciples, but I borrow it here so I can ask the evolutionist or he can ask me, "Who do you say that I (a human being) am?" The evolutionist tells me that I am simply a naked primate. I tell him that he is a creation of God. The evolutionist tells me that nothing really matters and we live, die, and that's it. I tell him that where you spend eternity is in the balance. God will hold us all accountable. I have Jesus as my defense lawyer. The evolutionist has

---

2. Gould, *Ever Since Darwin*, 161.

## Is Evolution Compatible with Christianity?

his own works, which are not even close to perfect. See why the uproar? See why the passion erupts? I certainly do, and the issue has nothing to do with science alone.

Recall from chapter 2 two types of science exist: *operational science* and *origins science*. Both deal with scientists interpreting clues as forensic detectives do. In these two fields, it is critical to understand that when one has ideas concerning origins, the ideas one has, in order to be considered valid, must not violate operational science. Next, they must be plausible or possible. Finally, and this is something not mentioned yet, they must have explanatory power, as Stephen Meyer states in his book, *Signature in the Cell*.

*What is meant by explanatory power?*

Meyer states what was said before, that there are differing types of science. He notes that Stephan J. Gould called origins sciences "historical sciences" and included ones such as geology and evolutionary biology. As these types were different from operational sciences like physics and chemistry, they would have to have different methods. Gould further stated that the historical sciences are testable, but not by the same means as the operational ones are. He said they are testable in that we can evaluate their explanatory power. This is a huge admission for the creation side. So if the origins science of the evolutionist seems to explain the facts, then it does not matter if it is true or not.[3]

So evolutionists often use tactics like Gould's in that they reason and figure in reverse and infer a history from its results that are seen in the present day. For example, they see fossils and other clues in the rocks today and infer a past ecosystem that would have caused the fossils to be present today. This allows them to reconstruct the Cambrian environment, even though they have never seen it.[4]

When one finds facts, how does worldview actually come into play? See Figure 2 below. This is the first of a series of diagrams from Answers in Genesis. The first one asks a very simple question concerning the two shapes seen. You are to draw what is missing. Another way to say it is: you have to draw what it looked like before. I have shown this diagram to my students, and they draw one of the possible sketches shown in Figure 3, or perhaps one not suggested. Now for the moment of truth. I ask them if they

---

3. Meyer, *Signature in Cell*, 150.
4. Morris, *Crucible of Creation*, 63–115; Gould, *Wonderful Life*.

would like to know what it really did look like before. Most desire to see what it was like. Look at the answer shown in Figure 4.

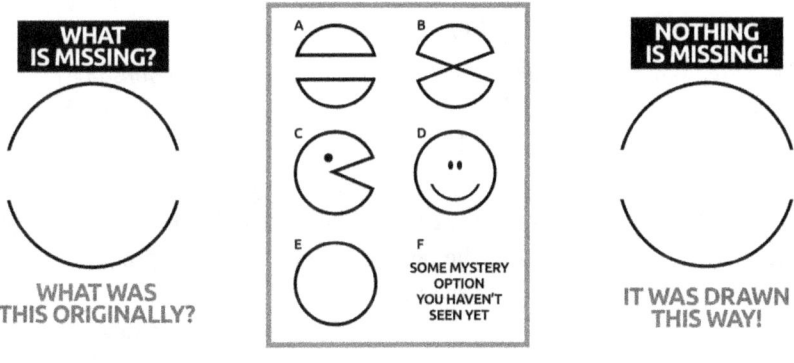

Figure 2.    Figure 3.    Figure 4.

After the first few years I did this, I got a response that brought me up short. Some actually accused me of cheating. I was taken aback. Why on earth would I be guilty of cheating? The students explained that they were told to draw what was missing, so they assumed that something was legitimately missing. I said that the students were not being cheated; they were being misled. They were not happy with that answer at all. I now had the teachable moment I was waiting for. I told them that this exercise is a perfect example of what I had been talking about that one's philosophy or worldview influences how to approach a problem, such as what conditions were like in the past when no one was there. Here is the sixty-four dollar question I asked the group: if a scientist's philosophy is wrong, will he ever get the right answer? The resounding chorus was a firm, "No."

Let us apply this to a very realistic example. As techniques improved, scientists began to actually read the bases in DNA. Soon genes were transcribed and functions found. Then they bumped into a wall of sorts. Stretches of DNA were found to do nothing. Due to the belief or worldview that Evolution Wrong is true, the term "junk DNA" was coined. These DNA portions were "obviously" leftover baggage from hundreds of millions of evolutionary years.

Some scientists studied them anyway. Lo and behold, they were not functionless as was thought. These DNA "trash" sections actually did have an extremely important function. We now know they code for RNA strands that are used to then make proteins or help with other cell processes. Thus

## Is Evolution Compatible with Christianity?

the evolution paradigm would have halted scientific progress. After all, who would want to waste time studying junk?

*How can a scientist gather facts, yet still get it wrong?*

Read Stephen Jay Gould's work in *Ever Since Darwin: Reflections in Natural History*. This book is a collection of essays written by the late evolutionist. One of his many underlined-by-me statements was this surprising one: "New facts, collected in the old ways under the guidance of old theories rarely led to any substantial revision of thought. Facts do not 'speak for themselves'; they are read in the light of theory. Creative thought, in science as much as in the arts, is the motor of changing opinion."[5] Note that the entire quote adds another problematic issue with interpretation. Gould says that "creative thought" is the force behind getting rid of old ideas or notions. So it is in how the interpretations are stated that is going to determine how correct the ideas are going to be.

This now readily explains the situation. Like the geologic facts that begged to be interpreted as evidence for plate tectonics, the "junk DNA" bases begged to have a use found for them. As the reigning "light" in power today, evolution has clouded the minds of those blinded by its light, thus they see the only interpretation available to them: that this is evidence for the idea of Evolution Wrong. Those not so blinded, however, kept working at the DNA bases and new knowledge was found. Now you can see that the evolution way of thinking actually held back scientific progress.

This is not the first time evolution has mislabeled facts. One can review the history of so called vestigial organs. At one time as many as one hundred human organs/structures were labeled as vestigial—structures that used to have a function in the evolutionary past, but now are reduced and/or useless. The list is now down to three: the appendix, the coccyx or tailbone, and wiggling one's ears. It really ought to be zero, because there are functions for the appendix and the coccyx (tailbone). Those who can wiggle their ears could be better interpreted as having a mutation similar to those who are double-jointed; it means nothing concerning evolution.

There is no one philosophy of science; however, there is one overriding principle that governs what evolution can espouse. This guiding philosophy is naturalism. Naturalism says that nature is all there is and that every explanation of phenomena must be a natural one. The supernatural is not applicable. As the supernatural is by definition outside of science,

---

5. Gould, *Ever Since Darwin*, 161.

anything attributed to it is by definition not science. This leads evolutionists to say that science says nothing about the supernatural or that it is neutral to religion. Remarks such as these are fine to make, but rarely do the popularizers of evolution walk the talk. If you encounter evolutionists, be ready for some of them to use unprofessional behavior when you voice your notion of creation by divine command as stated in the Bible. Note how "neutral" they act and what words they say!

Naturalism is okay to follow, up to a point. For example, our pagan ancestors explained creation, the cosmos origins, or even natural events like thunderstorms using mythological gods or powers. It was Thor or Zeus who threw the thunderbolts. Now we use operational science to explain thunderstorms. Thunder is the explosive expansion of air due to the extreme heat of a lightning bolt speeding to earth. We know that lightning is a discharge of excess static electricity from a cumulonimbus cloud. So everything has a natural explanation, they would say. I agree, but only up to a point.

Where naturalism falls very short as to be illogical is when evolutionists demand naturalism for phenomena that are clearly unnatural. As already stated, DNA is a coded set of instructions for protein synthesis. There is no science that shows that nature creates messages. It is incorrect to say that nature does create messages because DNA is natural and DNA has a message. Perhaps we need to define a message as that which carries information. DNA is a coded set of information steps leading to the construction of proteins. To illustrate the quandary the evolution side has with DNA, I relate this anecdote. I was at a science conference and, waiting for the presenter to begin, I struck up a conversation with the man next to me at the table. Finding out I was a creationist, he said, "You better watch out. I have my master's in organic chemistry." Instead of withering up and blowing away, I said, "Oh good. I've been wanting to bump into one of you guys for a long time." Suddenly he looked like a deer in headlights. "Tell me how DNA put itself together without the hand of God," I said.

There was dead silence from him. The presenter began, and so I wrote the gentleman a note. "Please tell me how DNA put itself together without the hand of God. Here is my address and phone number. If you cannot answer this, then I know that my side is right and yours is wrong." I gave it to him. Over thirty years later, I still await his response. For the price of a stamp he could have shot me down. The point is not who wins the debate, but who has the better explanation. It is true that one must be careful to

## Is Evolution Compatible with Christianity?

claim something for God that science cannot do. . . yet. For once science can do it, what will your faith in God stand upon? However, whenever science tries to duplicate what DNA or something in nature does, note how designed it is or how they cheat to get at the "natural" result!

*What are some examples of scientists cheating where evolution is concerned?*

A classic example of cheating is the famed Urey-Miller experiment. As you see in Figure 5, it consisted of glassware, spark wires, gas tubing, and water. The cheating came with some of the starter items and ended with the results not being correctly taught to students. First, the gasses used were assumed to be the same as the early earth. Modern origin theories say that volcanic gas was the source of the early atmosphere. Compare those spark jar gasses with volcanic gasses of today, and you will see that the list is very different from the list used in the apparatus. Therefore, the gasses used were chosen because they would make amino acids happen. If you know about cheating in cards, what they did was stack the deck.

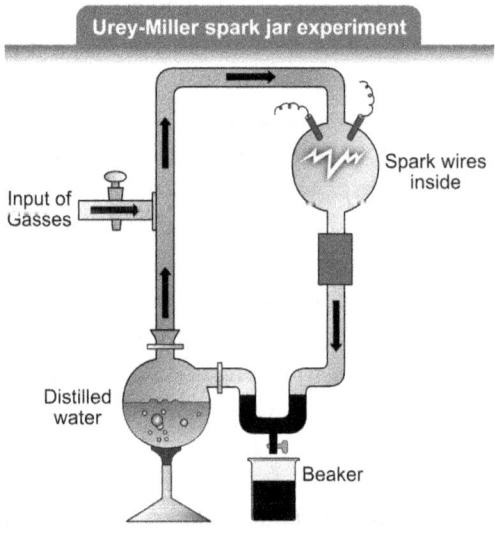

**Figure 5.**

Second, the device used items that represented nature's ingredients for amino acid production. The water used must be distilled water. This water is fine for any chemistry experiment, but this water was to represent the early ocean of earth. Anyone knows that water touching rocks will not be distilled for very long. Salts and other minerals would dissolve into the water making it not distilled at all. The reason they had to use distilled water is the experiment would fail if they did not use it.

Additionally, there is an issue with using spark wires. These stood for lightning. Since when is lightning a few dozen volts at best? Perhaps lightning evolved into the strong bolts of today? I think not. Another cheating item was the use of a beaker to keep the results separate from the rest of the experiment. Just where did that beaker come from four billion years ago?

The third and final area of cheating is that the results are not told correctly in sufficient detail. Were amino acids produced? Most certainly they were. Were other molecules, including garbage ones made? Of course they were. But most telling are the amino acids themselves. The molecules made were racemic, which means that some were right-handed and some were left-handed. To us who have both a left and a right hand, this difference does not seem to be very critical. Our cells, governed by biochemical laws, would have a lot to say to the contrary. See Figure 6 to help you visualize racemic amino acids.

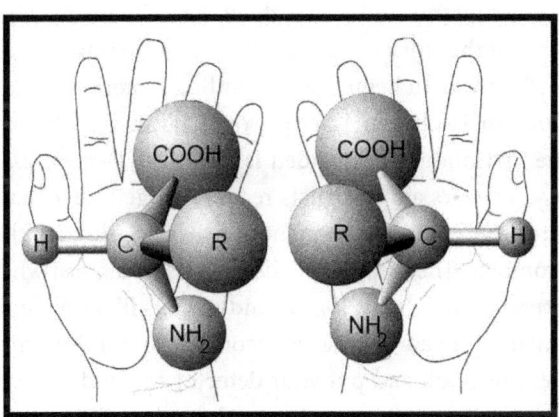

**Figure 6.**

*But how do these facts help the creationist side?*

It will take you finding another person, preferably one who thinks that nature is capable of creating a polypeptide chain without any guidance whatsoever. Put out your left hand. He does the same. Note that it feels a bit awkward, but it works as you shake hands. Now you hold the right hand straight out and have him "bond" his left hand with you. The palms are the bonding surfaces. Have him note that he must "warp" his arm position by turning it upside down in order to bond. Tell him that polypeptide chains must be folded in exactly the right shape or the whole protein is junk. Ask him to explain how the first protein formed with all the amino acids bonded in the correct position without a designing "hand" involved. Know that right- and left-handed amino acids bond together just as easily as the left-handed ones do and this is why the spark jar apparatus made biological junk.

*What does this have to do with evolution?*

Remember that there is no God allowed in materialistic/naturalistic science. Nobody is around to put proteins together properly. All proteins found in nature consist of left-handed amino acids only. What are the odds of only left-handed amino acids coming together in the proper order to make a protein without an intelligence guiding them? They are so small as to be zero for all practical purposes. So why aren't these facts allowed to be taught to high school and college students? Materialist philosophy currently reigns supreme and comes down harshly on any who would say, as the little boy of the story exclaimed, that the emperor is naked.

An excellent way to begin understanding materialist philosophy and its applications to Evolution Wrong would be to read books by Phillip Johnson. The first one recommended is *Defeating Darwinism by Opening Minds*. Johnson covers and instructs readers on getting a "baloney detector," which ironically was first thought of by Carl Sagan, another late great evolution promoter. The creationist's detector is to discern what is fact and what is interpretation, what is logical and what is illogical, and that which is experimentally proved to what is a con job of hand-waving or special pleading. Get this book and put your detector to good use whenever you face evolution "facts," readings, or those who hold evolution to be true.

Johnson also discusses various examples of what scientists ought to do and not to do. One item to not do is believe what you want simply because you do not like the alternative. One of the quotes used in my class is from a book review published in *Geotimes* September 1994 on p. 23. The review concerned the mass-extinction debate and how each scientist argued philosophy of science with no real effect on the end of the issue. The key sentence is the last. "A philosophical truth in the geological sciences seems to have been validated—we find what we look for."

The evolutionists find evidence for evolution and the creationists find evidence for creation. No surprise there. What is so surprising is that creationists play the same science game as they do, and they are accused of being unscientific. Illustrating this point is another experience I had. After my last presentation I made to our state science conference, an evolution-believing professor wrote my principal and accused me of several things. He wrote that I had abused the professional journals when I had this quote shown, among others. I wrote him back and responded to each accusation.

I asked him to tell me what the quotes meant when the scientists said what they said I said if they did not mean what they said I said they said. He wrote me back to simply invite me to a fossil dig he runs in the summer. I guess the scientists meant what I said they said. I did not abuse the journals.

Next we head into the material presented in high school biology texts starting with Darwin.

# Chapter 7

# Darwin Right and Darwin Wrong

"The difficulty is to detach the framework of fact—of absolute undeniable fact—from the embellishments of theorists and reporters."

—Sherlock Holmes, "Silver Blaze" (from *The Memoirs of Sherlock Holmes*)

Evolution as a paradigm rests upon three pillars. One is that life created itself, dealt with in chapter 11. The second is that there are long time spans of billions of years. Finally, the last pillar says that life forms can change into totally different life forms. Now we will deal with that instance of change. To do so we must begin with Charles Darwin, a familiar name to both evolutionists and creationists.

Darwin arguably did more to set in motion the creation/evolution conflict than anyone else in modern times. It was he who set the fires of evolutionism to flame. Had Darwin stated or examined only what was operational or proven science, he would have been on much safer ground. In fact, he would have corrected early creationists' error of thought, namely the immutability of the species. The phrase "immutability of the species" means the belief that God created animals and plants exactly the way they are today. In other words, animals do not ever change at all over time. The proper view today is that animals do indeed change, but over a very small range. It would be fairer to say that, as an idea, immutability of the species had days that were dwindling due to the work of creationists such as Edward Blyth. He and others like him were putting Darwin's natural selection, which said that nature chooses the best variations to survive, in its proper place, which is to be a conserver of traits and remover of variations unfit for survival.

*Just why is Darwin such a problem for biblical creationists?*

The main issue most creationists have with Darwin is that he overstepped his bounds. He not only sought to explain why species change, but also endeavored to explain the origin of life. The amount of change Darwin proposed possible in an animal or plant line was truly nothing short of utterly unbelievable. This latter error was due to his extrapolating small changes seen over short time into drastic changes ultimately over huge time spans. So did Darwin get anything else correct besides the fact that species can change, albeit in a very limited range? Of course he did. Let us examine his observations.

Most high school biology texts, including the one I used, cover these basic Darwinian tenets: (1) Overpopulation leads to a struggle for existence. (2) Variation causes adaptation to occur. (3) Survival of the fittest is at work, meaning those who are fit survive to reproduce. (4) Natural selection chooses the best variations to survive. (5) All life came from a common ancestor.

The first part of the initial tenet is true—there are places that struggle with overpopulation; the other part is open to interpretation. We know that there are many more offspring of any animal or plant than needed to replace the parents. But this does not set up competition. The reason for overproduction is simple. Other living things have to eat something, so even after they have eaten their fill, there will still be plenty of living things left over for reproduction. So in other words, the struggle for existence is really not to get eaten.

A cursory examination of any ecology chapter in a biology text will come up with a biomass pyramid. Producers or plants occupy the bottommost layer or level. Herbivores or plant-eaters fill the next level, followed by various consumers called carnivores. As one goes up the pyramid, each level gets smaller because there are fewer organisms in that level. Obvious to anyone who has seen any documentaries on lions, tigers, or bears can deduce that there are far fewer of the eaters than the eaten, no matter which levels one talks about.

The reason that this is so is due to the fact that most of the energy an organism takes in is used up in keeping the organism alive. There is also a small percentage that is lost as heat. But more important is that there is now less energy available to the higher level due to the lower level's energy use. So the struggle for life is to out produce those that get eaten so there will always be enough food for the higher levels.

*What about Darwin's second idea of variation? Isn't that true?*

Variations do occur and they do cause adaptations to happen. It is no stretch of the imagination to believe that from one fox type come the desert and arctic foxes. The colder environment favors heat-conserving adaptations, such as thick white fur, short legs, and short ears. Hotter environments favor the long ears for heat loss, longer limbs, and thin fur of a sandy color. The key question is: do these small changes in these animals, which are still 100 percent foxes, warrant a belief that animals can change into completely different animals? Not by any stretch of my imagination.

Imagination... a story hangs with that word. I once was accused of not having enough imagination by two evolutionist high school science teachers, which to them explained why I did not accept evolution. Then they walked away. Well, let's see now... I have written fifteen original plays as of this writing and four have been published. So much for no imagination. I should have responded, "I do not want imagination; I want information." I want to be shown with an experiment that if you change the DNA, you can get a totally new class of animal to appear instead of a mutated form of the original. At this point, a common evolutionist dodge is to state that I don't understand how evolution works—that you just do not get changes like that. Perhaps I understand not only how evolution works but also how it cannot work. The truth is I just do not accept illogical explanations and neither should you.

*So what is the source of variation?*

At this point we need to discuss the source of all this change. It is mutations, pure and simple. Without DNA altering, there would be no change at the visible level. Just how effective are mutations at changing things? As scientists study bacteria, they have found that there are some that have become resistant to our antibiotics. The reason is due to mutations happening within the bacterial chromosome. The type of mutation that usually happens is called a deletion. That means part of the gene controlling a certain aspect of that cell is lost. What it loses is the ability to absorb the antibiotic, so it is immune to that biochemical bacterial poison.

We are told that to understand this bacterial transformation, we need to believe that evolution is true. This is a total lie. One needs to understand bacteria, DNA, and how mutations work, not evolution. Besides, how can they claim that life forms became adapted and became genetically more complex—in other words, one bacterial chromosome becoming forty-six in humans—by deleting DNA? It is the height of illogic.

A key phrase I use with my students is that you can only change within a range. Change too much and you die. Evolutionists have done literally thousands of breeding experiments with *Drosophila melanogaster*, the humble fruit fly. We have fed them mutagens; we have irradiated them to bits. Do we get evolution? Not in the way they say it should go. All we have *ever* gotten is mutated fruit flies. Most mutants are less fit than the original "wild type" and would not survive outside the lab. It is highly proper to ask the evolutionist, "Where is the evidence for evolution?" and by that we do mean the macro kind which is the wrong kind.

*What about the notion of survival of the fittest?*

Obviously, if you are fit, you will reproduce, and thus your species will survive. This is where evolution does not really fit too well with what we do as humans. We constantly allow the "unfit" to reproduce. Look at how many people there are who wear glasses or contacts. What a drain on the human genome. Think how much money we would save if all those nearsighted mutants or flawed adaptations would just be allowed to struggle along with the rest of us without the aid others give them. Of course, I am being absurd.

*Is there such a thing as natural selection?*

Yes, indeed, there is. There is absolutely nothing wrong with saying so scientifically or scripturally. One of the most famous examples of natural selection used by textbooks is the peppered moth of England. Before the industrial era, there were lichens that grew on trees. Moths with a black and white mottled appearance were camouflaged and not readily eaten, whereas the solid black ones were more easily seen. The industrial era caused trees to be covered with soot from coal dust and thus black forms now had the camouflage advantage.

## Is Evolution Compatible with Christianity?

The problem in using this example is that after the population shifted from a peppered form to the mainly black variety, pollution controls ruled, lichens came back, and the peppered form was restored. Understand this: evolution went in reverse. And if it went in reverse today, how do we know it did not go into reverse yesterday? Other problems were said to exist with this classical piece of "evidence" for evolution, which is why it is not being used that much anymore in more modern high school biology books such as Ken Miller's 2012 version. Recent evidence does suggest that predation by birds and the coloration of the moths do indeed play a part in the shifting of a population's coloration.[1] It must be stressed, however, that this is no way points toward bacteria becoming basketball players. Thus the peppered moth data prove Evolution Right correct and Evolution Wrong as, well, just wrong.

*Where did Darwin get his ideas on evolution?*

What most do not know is that Darwin borrowed his idea for natural selection from others of his time. While that is not really a crime, most textbooks do not mention even a hint of this and credit Darwin with this idea. Darwin obtained this idea mostly from a creationist Edward Blyth, who wrote several articles twenty-two years before Darwin wrote his book. Darwin even lifted whole sections of Blyth's articles to write *On the Origin of Species*.[2]

Blyth's natural selection differed from Darwin's in several ways. First, Blyth said that this was God's way of preserving living things in their best-adapted form and weeding out the variations that would not make it. Additionally, natural selection does not create new features of a species; it only modifies existing features.

The question now is this: does natural selection cause evolution to happen? If you mean how to go from a common fox to the arctic and desert varieties, then yes. This means that Evolution Right is . . . right. If you mean natural selection causes a bunch of atoms to become atheists, then no. Natural selection merely acts to keep the basic form from becoming too different. Remember what happens if you change outside of the range. It also will be the environment that is doing the selecting. This means if you cannot adapt, you will die. But also recall that the adaptations must be

---

1. Purdom, "Peppered Moths."
2. Ham, *New Answers*, 274.

already present. If the warped gene for antibiotic resistance is not there, the germ will die and not pass on the resistance.

*But doesn't the dog-like mammal to whale sequence prove that environment can change animals?*

If this idea of the environment changing populations is true, then we need to start living in the water. After a few million years of doing this, aquatic humans will result. Now some will argue that this is nonsense, but the same thing happened to dog-like mammals and they became whales. Don't think so? Just ask Ken Miller or Joe Levine. (More on this in chapter 10 of this book.) That's what they imply in their biology textbook.[3] Those who do not think humans would become aquatic obviously must suffer from a lack of imagination.

To be fair, they do show a series of fossils organized by the assumption that evolution is true. The hind and forelimbs do suddenly change into paddles, and then the hind ones disappear totally. A modern porpoise or dolphin is shown at the end with a greatly reduced pelvic bone where the hips ought to be if it were a land mammal. The implication is that this bone no longer has any function, as it is a vestige or leftover of evolution. Nothing is further from the truth. In actuality, this bone is useful in the reproductive process as it provides an anchor for muscle attachment.

Conveniently skipped over are the actual facts that throw the changing of a dog-like mammal into today's blue whale into disarray. It is also a bit funny to read in *Finding Darwin's God* that Miller tells how Dr. Phillip Gingerich spells trouble for those against whale evolution as Gingerich has found not only one, but three additional in between species in the land mammal to whale line.[4] Yet when one watches the DVD series *Evolution: The Grand Experiment* and listens to Dr. Gingerich's comments on those very same in between stages, he sites problems with them.

*What are some examples of those difficulties for whale evolution?*

If one uses fossils, then you end up with a diagram like the one found in the textbook I teach from. However, if you use DNA, like scientists at the Tokyo Institute of Technology, then hippo DNA is the closest match to today's whales.[5] So which is it: hippos or a dog-like carnivore as the ancestor of the

---

3. Miller and Levine, *Biology*, 466–67.
4. Miller, *Finding Darwin's God*, 264.
5. Monastersky, "Whales' Tale."

whale? The ancestor *Ambulocetus* has another problem being an ancestor of today's whales—its eyes. Dr. Phil Gingerich, a paleontologist at the University of Michigan, states, "Ambulocetus has its eyes raised on top of its head in a very strange way, and it is unusually large for an early whale . . . maybe it's not on the main line [in whale evolution]."[6] Finally, there is the problem of the missing tail for the *Rodhocetus* fossil. The end of the tail is missing, thus we have no idea if it had flukes or not. Gingerich commented on this: "I had speculated that it might have had a fluke . . . I now doubt that *Rodhocetus* would have had a fluked tail."[7] What a fluke of bad luck for Miller-Levine and our biology text. More about this will be shared in the fossils section.

*Do we truly all have a common ancestor?*

The last item Darwin proposed was that all life forms have a common ancestor. This is supposedly proven by the concept of homology. Figure 7 shows the evidence that supports this notion. Diagrams using this idea usually show a prehistoric lobe-finned fish and detail the bones that are actually found in those types of fins. The bones are sometimes color-coordinated to match the ones found in other vertebrates. What is even sometimes just as amusing is that some authors stress that the bones have the same names.

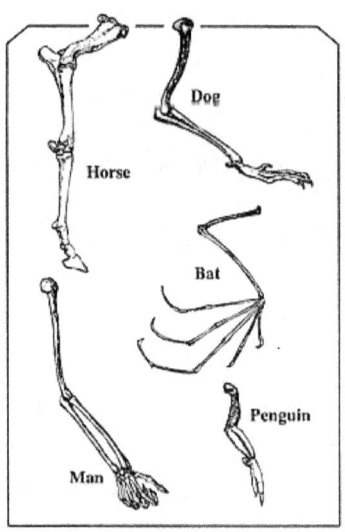

**Figure 7.**

---

6. Werner, *Evolution*.

7. Werner, *Evolution*.

What is true is that this evidence just as easily supports the idea of a common designer as well. Think of your favorite band or artist, poet, etc. Were the person/group to create a brand new work of art; and it were mixed in with three other works by different people, would you be able to pick it out? Of course you would, as you are familiar with the artist's creative genius signs. Why would we limit God to having to create such a hugely varied creation, making every animal radically different from others of the same order or family level? That would be horribly confusing to study. After all, with apologies to Dawkins, it only *looks* like it evolved, when it was really created.

As stated before, another problem with homology is the fact that the evolutionist stresses that the bones have the same names. Well, guess who gave them names? There is an Abe Lincoln story applicable to this issue. Abraham Lincoln debated his opponent regarding slavery. His foe wanted slaves to be called property. Lincoln asked him, "If I call a tail a leg, how many legs does a cow have?" "Five," was the response. "No, it is still four. Calling a tail a leg does not make it one." So calling a bone in a bird wing an ulna does not make it one.

A final flaw with this issue is that this idea of a common ancestor is proved by homology. But homology is proved by the fact that there is a common ancestor. This is circular reasoning. Figure 8 shows how this works, or doesn't work, as the case may be. One cannot logically use an item for proof of a second item, and then turn around and use that second item to prove the first. It is like trying to pull yourself up by your own hand. What is needed is a third source of proof that is outside the system.

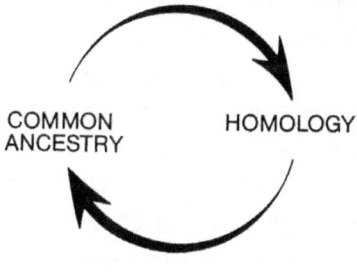

**Figure 8.**

Evolutionists prop up the homology concept using other evidences. There have been attempts to use biological molecules, such as cytochrome C. It is true that one can construct a tree showing a progression of cytochrome

C differences between the lower animals becoming the higher ones. What is actually the case is that these "evidences" are carefully picked to show what the evolutionists want to show.[8] There are animals which would totally scramble the so-called order of evolution were we to randomly choose living things to test.

*What is it about Darwin that causes so much uproar?*

Benjamin Wiker examined just such a question in his book, *The Darwin Myth*. For one, Darwin's lie was that "he could have his moral cake and eat it too, pushing forward a godless account of evolution that made morality a mere transient effect of natural selection, and at the same time holding up particular moral traits, such as sympathy, as if they were somehow the aim of aimless evolution."[9] Wiker continues to explain another lie or myth, which Darwin himself helped to continue. This was that there is an either/or choice and nothing in between. Either one has godless nature with evolution as its creative force or one has young earth creationism with God creating every single finch species or ant species individually.[10]

Wiker, at the end of his book, appeals to us to not be rigid literalists and thus demean the grandeur of evolution. To him, and Dr. Ken Miller, compromise is the road to the promised land of no more fighting. It may be the road to peace, but it is more important to ask if it is the road to truth. Good question, but one that gives secular mankind fits. Even Pontius Pilate asked Jesus, "What is truth?" (John 18:38). Christians have a definition provided by Jesus Christ in John 14:6, where he said, "I am the way, and the truth, and the life." Note that the Greek does not say "a" way but it has him saying "the" way. This applies to the other words as well. Jesus is claiming to be the absolute truth or Truth, if you will.

Money and/or power do not mix with truth, especially if truth says those in power are wrong or lying. That money drives much in science, even evolutionary "research," is revealed in a conversation between Dr. Lynn Margulis and Dr. Richard Lewontin. He had just lectured on how he had mathematized a large variety of Darwinian data such as random mutations, sexual selection, etc. He later confided to Dr. Margulis, "You know, we've tried to test these ideas in the field and the lab, and there are no measurements that match the quantities I've told you about." This appalled

8. Luskin, "Problem 6."
9. Wiker, *Darwin Myth*, 149.
10. Wiker, *Darwin Myth*, 138.

## Darwin Right and Darwin Wrong

Dr. Margulis. Dr. Margulis said, "Richard Lewontin, you are a great lecturer to have the courage to say it's gotten you nowhere. But why do you continue to do the work?" And he looked around and said, "It's the only thing I know how to do, and if I don't do it, I won't get my grant money."[11]

I think it is time for the American people to ask just what grant money is actually being spent on so it can be used wisely. In my opinion, we need to find cures for cancer or better solar energy collectors than to be spending it on researching evidences that lead nowhere. I do find it amazing that the truth has finally been put into print, but who will be moved to ask some tough questions about it? How about you?

*Darwin's ideas are based upon the philosophical idea of materialism, which claims that only what is seen and observed is real. Is it true that this materialism is responsible in part for some of the dictators of the world to arise?*

The sad thing is that Darwin's own push for materialism allowed others to use his idea as a basis for their power grabs. Modern examples would be the ever-mentioned trio of Hitler, Stalin, and Mao. True enough, Christians have blood on their hands, but these three make Christian sanctioned wars a Sunday afternoon picnic. It should give us pause here to ask which other idea in science has caused so much grief? Protein synthesis? Mitosis? Creation? Intelligent design? Agreed that the last two are hardly scientific ideas, but none of these listed tells us we do not matter. Evolution, as an idea, tells us that nothing matters.

Coupled with evolution is one of the two ways the end of things will come. The first is that everything will eventually end in universal heat death. This is the idea that the universe will eventually run out of energy, and then everything is at the same temperature, which means all processes come to a halt. The second idea is called the Big Crunch. This means that gravity eventually wins and every galaxy, star system, you name it, eventually collapses into the original super dense particle that started it all.

With either scenario it basically tells you to do what you want, as there is no god to answer to and in the end, we are all dust anyway. "There is no remembrance of former things, nor will there be and remembrance of later things yet to be among those who come after" (Ecc 1:11) Solomon's words from Ecclesiastes 1:1–11 provide an insightful look into the end result of materialistic evolution in the manner of Dawkins and others. The RSV version speaks of meaninglessness.

11. Teresi, "Discover Interview."

The truth is that creation, and Christianity, provide just the opposite. Consider that what we see now will be destroyed for sure, but not by gravity. Second Peter 3:12 tells us that the heavenly bodies will melt as they burn. The old order of things must be destroyed so that the new may be created. Second Peter 3:13 states that "we are waiting for new heavens and a new earth in which righteousness dwells." In Matthew 6:19 Jesus tells us to store up treasures in heaven. Finally consider that our heavenly dwelling will be beyond our comprehension. Read the description of the heavenly city found in Revelation 21. All sorts of jewels adorn the city. Gold is used to make the roads. But the greatest reward will be that we will be with Jesus forever. Revelation 21:4 says it best, "He will wipe away every tear from their eyes, and death shall be no more, neither shall there be mourning, not crying nor pain anymore, for the former things have passed away."

# Chapter 8

# Time is of the Essence

> "I'm afraid I give myself away when I explain. Results without causes are much more impressive."
>
> —Sherlock Holmes, "The Stockbrokers Clerk"

It can be said that evolutionists have a trinity just as biblical creationists do. The three "persons" of this scientific religion's trinity are matter, chance, and time. The third member of this trio deserves its own prime spot, or prime "time." There are evolutionists who refer to time as the hero of the story. Without time, and lots of it, the evolutionary story is quite impossible. This quote says it all:

> Time is in fact the hero of the plot. The time with which we have to deal is on the order of two billion years. What we regard as impossible on the basis of human experience is meaningless here. Given so much time, the "impossible" becomes possible, the possible probable, and the probable virtually certain. One has to wait: time itself performs the miracles.[1]

Creationists hold that time is not the hero at all. In fact, based upon evidence seen in the laboratory, it does not matter how much time is involved. Impossible stays impossible. Biblical creationists hold that the earth

---

1. Wald, "Origin of Life," 48.

has to be at least six thousand years old. This is based upon biblical chronologies using the ages of how long pre-flood people lived as the book of Genesis clearly states.

*What does it mean that the age of the earth is one of the pillars or supports for evolution?*

It was previously stated that the idea of long time spans, like billions of years, is one of the three main supports or pillars of evolution. If the above quote is not evidence enough, then consider this written in *The American Biology Teacher* (*TABT*) by Rosanne W. Fortner: "An understanding of organic evolution rests on a fundamental awareness of 'deep time,' describing the great age of the Earth as revealed through its geologic structure."[2] If item A rests upon item B, then item B supports A. So the pillar idea is not as farfetched as it seems.

*How does Fortner state her case?*

Fortner continues in that same article that deep time is critically tied to evolution. "One of the major obstacles to acceptance of evolution as a valid scientific concept is a lack of understanding of the age of the Earth."[3] She continues on with James Hutton and how early scientists held that the earth was only about six thousand years old. The section's last sentences are indeed telling. "In accepting the Earth as being of great age one can reject the idea that the world was created specifically for human use. We now understand the planet evolved over billions years, and the human species is a very recent result of biological evolution. Because some species have become extinct along the way, there is reason enough to believe that we may not be the ultimate and culminating product of the evolutionary process!"[4]

Several comments are warranted concerning the ideas presented. She honestly admits that without billions of years, the whole edifice of evolution is in trouble. Without their "god" to do the miracles of lining up all left-handed amino acids to make the first protein, to have DNA nucleotides randomly try to create information, for countless fortuitous mutations to pile up and not kill the mutants, the natural explanation of everything collapses. The question then needs to be asked—why does the evolution side cling so tightly to Evolution Wrong? If the house of cards comes tumbling

2. Fortner, "Down to Earth," 76.
3. Fortner, "Down to Earth," 76.
4. Fortner, "Down to Earth," 76.

down, we will, according to the evolution side, all be headed for a new time of Dark Ages where science dies away and mankind suffers. Too bad most evolutionists do not study the history of science. Were it not for Western creationist Christians, who were the first scientists, science, as we know it, would not even exist.

Fortner says that we can reject the idea that the earth was made solely for our use if it had evolved over billions of years. I do not follow the chain of reasoning here. What does time have to do with purpose? Does the house she lives in stop being her house after so many years? Were she to live to five hundred years old, and she made me a shovel when she was in her thirties, would it stop being my shovel simply because she is old? To know that God formed this Earth to be inhabited and filled with beings like us, one can read several Bible passages. One of the best is Isaiah 45:18 ESV, "For thus says the Lord, who created the heavens (he is God!), who formed the earth and made it (he established it; he did not create it empty, he formed it to be inhabited!): 'I am the Lord, and there is no other.'" What is also plainly stated is that God is the Creator of this world of ours. There is no hint of evolution going on at all. Now it is true that this verse does not specifically tell us how long it took for God to do his creating, but we do know that this passage perfectly mirrors what Genesis says in that God did it without needing anything at all.

*Have humans fully evolved?*

The last point Fortner makes is a most interesting one. If evolution is true, then we are not the end of the line. We are simply the "amoeba" as the starter life form for the next two billion years of evolution. We hold no special place at all. I have even heard of some saying that humans are a pest species on the planet, due to our abuse of nature. If we are pests, then shouldn't we be exterminated? If we were to become extinct, would it matter? Fortner seems to say that it would not matter at all. The universe would not care a whit were we to blink out of existence. Interestingly she claims that only "some species have become extinct along the way," yet when one reads other evolution scientists and book authors, they routinely claim that over 90 percent of the species that have ever lived have gone extinct.[5]

What is fun for any creationist is to say this bit of "nothing special about us" philosophy to an evolutionist and see how they respond. I did this while emailing an evolutionist in 2010, and he said I sounded depressed

---

5. National Geographic, "Mass Instinction."

when I rambled a bit about this lack of purpose in life. I was not depressed at all. I merely stated his worldview consequences right back at him.

*How has science proved the age of the earth?*

First, we need to cover how the earth started to age right before our eyes. Miller and Levine state that it was James Hutton who began the modern ideas of geological history. Hutton observed that geologic processes seemed to be related to geological features. As lava cooled, it formed the igneous rock basalt. Other rocks appeared to be mixtures of parts of smaller rocks, like conglomerates. These jumbles resembled the washes at the feet of mountain streams.

Hutton further proposed that some rock layers appear to have been tilted up by tremendous earth forces working over long periods of time to create mountains. Mountains become sediments as they are slowly worn down and eroded to be deposited elsewhere. Seeing all these facts caused him to believe that the earth was much more than the six thousand years of Bishop Usher fame. So Hutton was the first to consider the concept of deep time. And as the old commercial went, when Hutton spoke, others listened.

Charles Lyell continued where Hutton left off. Lyell introduced the concept of uniformitarianism. This means that processes of the past work like they do today, and so the present is the key to the past. Notice that the root word of this long term is *uniform*, meaning the same. Not only do the same processes happen, but they also happen at the same rates. Thus if a process deposits a few millimeters of soil in a stream bed, then it would take many millions of years indeed to pile up to a mountain.

Lyell also noted that rivers cut into bedrock to make stream channels. Rivers slowly cut deeper and deeper into the layers. Lyell argued that earth could not be merely a few thousand years old, because there wouldn't be enough time for a small river to carve such a ravine into the rocks.

Miller and Levine mention that it was Lyell's work that mainly convinced Darwin that there was enough time for the gradual accumulations of variations to allow evolution to occur. The irony is that Lyell vehemently denied what he called transmutation. In his second volume, Lyell states that Lamarck confused variation within a species with unlimited variation that Lyell said was impossible. Lyell even pointed out that breeders, no matter how hard they try, simply cannot transmute any living thing into another different animal or plant.[6] It is strange that neither Levine nor Miller

---

6. Wiker, *Darwin Myth*, 38.

mentions this aspect of Lyell at all. To me this is just one more example of evolutionists cherry-picking their facts so students will not be in danger of hearing the truth.

What Hutton and Lyell refused to consider was that floodwaters can be tremendous carvers and agents of erosion. In fact, many geologists today are returning to the concept called *catastrophism*. This term means that earth sometimes has violently powerful events that cause the landscape to be changed. Witness how Japan was affected geologically by the recent tsunami and earthquake. Too bad these two did not consider that the world's greatest gorge, the Grand Canyon, could have been formed by a lot of water in a little time.

*Is there evidence that can be better interpreted as having a catastrophic origin?*

Research the Scablands of Oregon and Washington. Even secular geologists assert that when the last Ice Age created glacial Lake Missoula there was some sort of a dam holding back vast amounts of water. One day the dam broke, releasing a torrent of water. The resulting erosive forces created the Scablands we see today in a matter of days, and not millions of years.[7] Thus one cannot judge age by appearances.

*Is there another way to date an object or landscape other than by what we can see?*

Archeologists use a dating technique using radioactive carbon, an isotope of normal carbon. This dating method, while fairly accurate with archeological finds, such as Egyptian mummies and the like, cannot be used to date rocks. It is only valid on organic finds and only those within the time span of about fifty thousand years. This is due to the short half-life of the isotope.

*How does radiocarbon dating work?*

It is known that plants take in carbon dioxide ($CO_2$) and water to make glucose. Most of the carbon in $CO_2$ is the normal type of carbon, which is carbon-12. Occasionally, an isotope of carbon forms in nature. This one has more neutrons in it than regular carbon. The radioactive form is known as radiocarbon, or carbon-14. Therefore, plants take in this form along with the regular form of carbon. So this radiocarbon can be found in living and

---

7. Ham, *New Answers Book*, 218.

dead forms of plants. Not surprisingly, the carbon-14 form gets into animals and us because we eat the plants or eat the animals that eat the plants.

Once an animal or plant dies, it no longer takes in the radiocarbon. The amount of carbon-14 will become less as time goes on due to radioactive decay. So if we can determine the amount of radiocarbon in an organic remain and then compare that to what is in the atmosphere today, then we can calculate how old the item is based on the known decay rate.

*Are there any problems with this dating method?*

Yes, there are. No creationist denies the age of the mummies or other historical artifacts dated by this method, as one can find other items that corroborate the age found. It gets a little funny, though, when fossils or remains are found that are from the time before the flood, and so are only about four thousand years old. Fossils are discussed in the next chapter, but will be mentioned a bit here, as there have been some amazing discoveries with fossils lately.

Back to the radiocarbon dating method problems. The main issue or problem is with assumptions. The first assumption is that radiocarbon is produced and decays at a certain rate. Right now there exists a ratio of carbon-14 to carbon-12 of one atom to one trillion atoms respectively. Dr. Willard Libby, the discoverer of the radiocarbon dating method, assumed this ratio to be consistent or at equilibrium at present. He based this assumption upon the "known" age of the earth as billions of years old.

He actually did find out that there is not a state of balance or equilibrium between the two carbon isotopes. Originally, he calculated that it would take about thirty thousand years for this equilibrium to build up, starting with an earth that had no carbon-14 at the start.[8] Libby decided to ignore the unbalanced isotopes and chalked it up to experimental error. Curt Sewell has indeed documented that for our modern era, the discrepancy stands as follows: "The Specific Production Rate (SPR) of C-14 is known to be 18.8 atoms per gram of total carbon per minute. The Specific Decay Rate (SDR) is known to be only 16.1 disintegrations per gram per minute."[9] Because this shows that the ratio is not yet at equilibrium, and Libby stated that it takes about thirty thousand years for the equilibrium to be established, then this suggests that the earth might not be billions of years old.

8. Libby, *Radiocarbon Dating*, 8.
9. Sewell, "Carbon-14."

The second assumption is that the amounts of carbon-14 have remained constant over time. But science is based upon data and measurements. Can they prove it has always been like this? No. What if the amount of carbon-14 was actually higher than it is today? Most evolutionists will tell you that the earth's climate has undergone large changes over its billions of years of history. Most will tell you that during the reign of the dinosaurs, the earth had a much warmer climate. One way to make it warmer was to have more carbon dioxide in the air. Again, without actually traveling back into time and measuring this, they have no way of knowing.

The final problem is more of a challenge to the evolution-believing scientists. It is now known that organic material can be found inside fossilized dinosaur bones. One wonders why they have not dated this organic material using carbon-14 dating methods. They might respond that it would be a waste of money, as none would exist. But again, without actually testing for it and not finding it, *how* do they know? One might ask what they are afraid of finding if they do test it.

*Has anyone actually tested supposedly old carbon based material to see if any radiocarbon remains in it?*

There are findings that cause serious challenges to the evolution interpretations. The RATE (Radioisotopes and the Age of The Earth) team investigated such findings. There are coal deposits that have been assigned to different evolutionary ages like Mesozoic or Paleozoic. The RATE team obtained ten samples spanning the various eras from the U.S. Department of Energy Coal Sample Bank at Pennsylvania State University. Special care was taken to avoid any contamination. The samples were sent to one of the foremost carbon-14 labs in the United States. In all cases, the lab found measurable amounts of carbon-14 in each of the samples.[10] According to evolutionary ages, there shouldn't be any radiocarbon in any samples because they are too old. Do these facts tell you something? Mere hand waving explanations just won't do. As in the story of the emperor's new clothes, the idea that Evolution Wrong is science and critical to anyone doing science is that emperor. Biblical creationists can see that the emperor is naked.

The RATE team also investigated radioactive minerals that are used to find the ages of rocks without any organic remains in them. Such rocks are often dated in the billions of years. Studies covered the retention of helium in zircon crystals, discordant dates from the Grand Canyon layers,

10. DeYoung, *Thousands*, 53.

and radio halos found in granites of great age. Andrew Snelling is one creationist geologist who worked on the rate team. He wrote three chapters in the book summarizing the RATE team's data and conclusions in *Thousands . . . Not Billions*. I have chosen him for mainly one reason. The evolutionist teacher I was e-mailing tried to show that Dr. Snelling was not a creationist, but had written in the past as if evolution was true. I sent the paragraph to Dr. Snelling and he replied that this quote was his, but was taken entirely out of context, and that he had explained the fallacy of this red herring years ago.

*So how are inorganic fossils dated?*

The main method used to date fossils is an indirect one. Recall that radiocarbon is used if the bone has not yet turned to stone. As long as some organic part is present in the remains, then the previous dating method can be applied. If the bones are fully mineralized, then radioactive dating methods are used. There are several of them. The more common are uranium becoming lead, potassium becoming argon, and rubidium becoming strontium. The dating techniques are used to discern the ages of igneous rocks near the layers with fossils in them. Then a guess is used to estimate the age of the layers containing the fossils in question.

This area of rock dating has its own issues. There are three assumptions that must be made in order to even hope to arrive at the true age of the rock layer. These are listed in a previous earth science text I used. See if you can figure out the flawed logic of trying to determine a rock's age based upon these three assumptions: (1) None of the lead escapes the mineral. (2) No outside lead is added. (3) No lead from a non-radioactive source was present to begin with.[11] There are other assumptions used to arrive at the rock's age, such as that the decay rate has been constant over time.

Hopefully you can see that there is no way any geologist can tell if any of these are true, therefore the ages determined for any rocks are just guessing games. Why don't Miller and Levine tell high school students these facts? Obviously, it is because were the truth known, then the whole evolution idea vaporizes. This list of assumptions was presented by me at one of my first talks at my state science conference. An attendee angrily asked me what creationist tract I got that "junk" from. When I told him it was the American Geological Institute's textbook, he stopped talking instantly. The only slam he could think of was, "Well, I bet it isn't on the state approved

---

11. American Geological Institute, *Investigating Earth*, 307.

# Time is of the Essence

list." Storming off, I wondered if he realized what he was saying. The state will approve what you teach? What if the state were creationist, would he still teach it?

*Do all Christians hold to the original six-day creation or that the original days mentioned in Genesis were twenty-four hours long?*

Another major issue surrounding this debate is sometimes between Christians. What does the word *day* mean in Genesis? Christians interpret this word in two different ways, often depending on which denomination they are. More liberal Christians tend to view the word *day* meaning a long time span, while more conservative Christians say that the word signifies a twenty-four-hour day. The reason behind most liberal interpretations center around the notion that Genesis 1 and 2 are poetic in nature and therefore are not meant to imply anything literally true.

The problem with calling Genesis a poem is that the verb form is all wrong. Stephan Boyd, a biblical Hebrew scholar, states that the finite verb form known as *preterit* was studied. He compared biblical events that were told using both poetic and narrative styles. Find the exodus story from Exodus 14. Compare that with Moses' song in Exodus 15:1–19; the same event is told in two different ways. The preterit verb forms are different.

In all Old Testament accounts of real time events, such as a battle or crowning of a king, there are a high amount of finite verbs of the *narrative* type. For actual poems, such as Psalms or Proverbs, there are a high amount of finite verbs of the *poetic* type. Each style is not 100 percent one form of verb or another, but a mix occurs. Narrative texts have a ratio of around a median value of .52 (total preterits to total finite verbs) and the poetry ones have a median value of .04. The Genesis account has a value of about .65.[12] The facts call for a reasonable conclusion that Genesis is written as narrative or a historical document.

There are two other facts conveniently ignored by those who wish to twist God's word to convey a meaning never intended by the stenographer of God's inspired message. First, it is crucial to note that there are two sets of words always used when the days are mentioned in Genesis 1. They are "evening and morning" and a number word like first, second, etc. Hebrew scholar Dr. James Barr, who does not hold Genesis 1 to be true history, tells us, "So far as I know, there is no professor of Hebrew or Old Testament at any world-class university who does not believe the writer(s) of Genesis

---

12. DeYoung, *Thousands*, 166–67.

1–11 intended to convey to their readers the ideas that: a) creation took place in a series of days which were the same days of 24 hours we know experience."[13]

In addition to the above quote, we can do a word count of the word *day* in association with either a number word or the words *evening* or *morning*. We come up with this tally: it is used 410 times and in each case always means a twenty-four-hour day, and it is used with evening or morning twenty-three times and also means a twenty-four-hour day in each case. So there you have it. The rest of the Old Testament uses it to mean a regular day. Why is Genesis 1 exempt? Is it so fallible man's science, which is fact today and false tomorrow, may be held *above* God's word? If this is logical, then we Christians are indeed insane.

The second fact clearing the way for the intended meaning of the word *day* is found in Exodus 31:12. This verse is the Third Commandment for many Christians. Read what it says. God tells us to work six days and take a day off for relationships with family and God, because he worked six days and took a day off. That is really something when one thinks about it. We do not have a Father who tells us to do as I say, but not as I do. We have a God who could have created the universe in six pico-seconds, but chose to slow down as he already foreknew the commandment that was coming down the line. So wanting to be consistent and follow his own rules, he did it that way so we would do so for our weeks.

You, the reader, ought to realize that he follows his own rules no matter what. He demands a blood sacrifice for sin. He demands that the sacrifice be perfect. His total love needs to meet his total justice. Jesus fills the bill, and we can thank God for that, because there is *no* other way to get to heaven.

---

13. Batten, *Revised and Expanded*, 39.

# Chapter 9

# Fossils as Evidence for Which Side?

> "Circumstantial evidence is a very tricky thing.... It may seem to point very straight to one thing, but if you shift your point of view a little, you may find it pointing in an equally uncompromising manner to something entirely different."
>
> —Sherlock Holmes, "The Boscombe Valley Mystery"

This chapter will begin a critical section dealing with the evidence for evolution presented in textbooks. Subsequent chapters will cover each set of evidences that are related to one another. One of the most common evidences used to prop up Evolution Wrong is fossils. Whenever the evolutionists bring up fossils as a proof of evolution, one ought to respond, "I have a bone to pick with you."

*What exactly are fossils?*

Fossils are remains or traces of past life. The remains can be parts or whole organisms that have been preserved in rocks or other material. Traces are something left by an animal or plant. One example of traces is tracks or tubes found in sedimentary rocks. Impressions or imprints of leaves found in rocks are another example of traces.

*How are fossils made?*

Several types of fossilization processes have been known to happen. The more common would be petrifaction, where silicate minerals replace the organic part. This literally turns the bone to stone. Often called permineralization, it occurs after sediments bury an animal having bones or hard parts, like a shell. The soft organic parts decay, leaving the hard parts behind. These shells or bones decay as well and the minerals of the sedimentary rock replace the organic bones or shells.

Another well-known fossilization process is freezing, such as what happened to Siberian mammoths. But the most famous process, shown in the movie *Jurassic Park*, involves amber. Amber is fossilized pinesap, in which a small animal or insect can be trapped in; as the pinesap hardens, the animal or insect becomes fossilized. Interestingly, this makes the bug a fossil inside a fossil!

*What is needed to make a fossil besides the dead animal or plant?*

There are generally four items needed to make a fossil. The first is rapid burial. This is a must, because scavengers make short work of any carcass or plant material. It is this fact that conveniently explains why so few fossils form today. The second is material to do the burying. As many fossils are found in sedimentary rock, it is a given that sediments of various types ought to do the burying. For example, one needs sand to make sandstone, muds to make shale, etc.

The next two items needed are water and pressure. Have you ever seen anyone mixing a bag of cement with water? Almost the same process was at work to create the various types of sedimentary rocks we see. Interesting to note is that it does not take a long time to make cement. You might even say that cement is a concrete example of why rock formation does not take millions of years. Pressure is needed to compress the rock layers, along with the animals inside them. This is why many fossil forms, like fish, are flatter than those in real life.

*Is there an event that gives us all four items?*

Indeed there is. The great flood described in Genesis 6–9 easily explains many fossil facts we find today, many of which are quite problematic for the evolutionist or the Christian who believes that God took many years to create the earth. Consider that fossils are found worldwide, and you can match that with the flood being global in scope. The various sedimentary layers are easily explained in that the tides still worked four times per day.

## Fossils as Evidence for Which Side?

Consider also that fossils follow the rule of 95. It means this: 95 percent of all fossils are shallow ocean life, such as corals and shellfish; 95 percent of those left are algae and plants; 95 percent of this small fraction is invertebrates, such as insects. The remaining 0.0125 percent are vertebrates, of which 95 percent are from the Ice Age after the flood.[1] Later on in this chapter this will be dealt with in more detail.

*According to many biology textbooks, fossils are supposedly great evidence that living things have changed over the course of time.*[2]

Fossils are supposed to show the progression of life from the lower, less complex forms to the higher forms of greater complexity, although the tune now is to say that all forms are just as complex as all other forms. Basically, fossils are said to be a record of how life has changed over time. It is true that fossils are the only things we have to help us know what life in the past looked like. While we can learn a great deal about that life from just the fossils alone, they also provide a springboard for speculations about other areas of that life. Remember that it is not wrong for a scientist to speculate. That said, it is crucial that you also remember that his speculations are biased by one's worldview, and that those speculations are simply just guesses with no direct experimental support.

*How have high school biology textbooks dealt with fossils over the years?*

What one finds fascinating is how a textbook will change from one edition to the next, even though it is written by the same author(s). Such is the case with a typical high school biology text. Let us compare Ken Miller's and Joe Levine's *Biology* textbook. One edition was published in 2002 and another one in 2012. Read what the authors had to say about fossils in 2002: "The fossil record provides evidence about the history of life on Earth. It also shows how different groups of organisms have changed over time."

Now compare to what they say in 2012: "From the fossil record, paleontologists learn about the structure of ancient organisms, their environment, and the ways in which they lived."[3] Lest some critic accuses me of playing fast and loose with the quotes, he can look for himself. Both sections come from the heading "Fossils and Ancient Life." They are also

---

1. Snelling, "Where?"; Morris, *Young Earth*, 71.
2. Miller and Levine, *Biology*, 2002 edition, Section 17–21.
3. Miller and Levine, *Biology*, 2012 edition, 539.

identified as important with a key shaped icon. Note that both are discussing the fossil record.

What is also very interesting is that the newer version has totally dropped the idea of fossils showing change over time. They also drop the notion of fossils being a record of history, as in like the geologic column. Another change made by the two authors is that in 2002 they state that the fossil record shows that over 99 percent of the animals that have ever lived have gone extinct. But in 2012, that fact has become extinct itself. Perhaps they had to drop it, as sharp students would be asking why we are so concerned about preserving biodiversity when nature keeps wiping out so many life forms.

Miller and Levine ignore modern ideas of fossilization, ideas that are encompassing catastrophism more and more.[4] How fossils form is given the standard slow deposition treatment found in antiquated textbooks. The diagrams of Figure 9a–9c show how Miller and Levine explain this process.

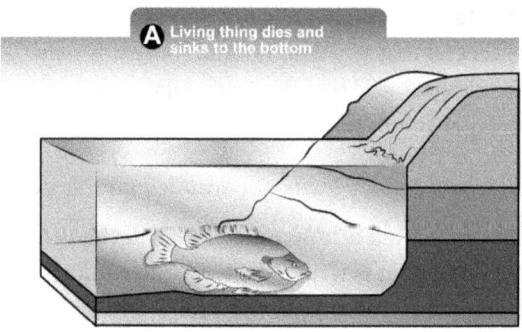

Figure 9a.

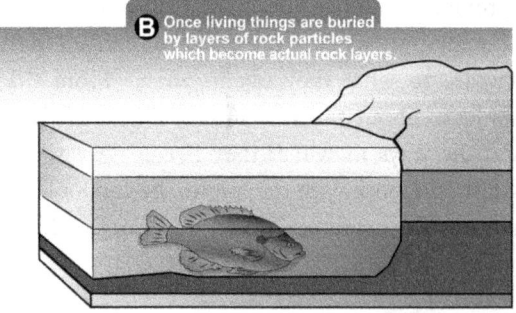

Figure 9b.

4. Miller and Levine, *Biology*, 465.

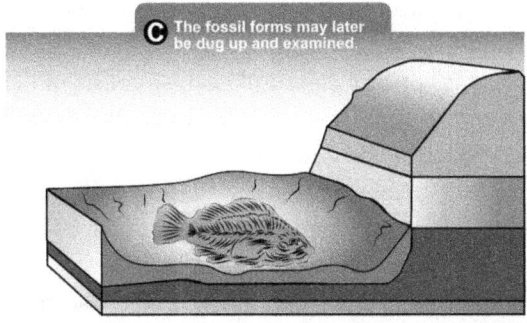

**Figure 9c.**

These diagrams strongly suggest slow burial near some river mouth. A fish may just get buried, but one would need a flood of rapid and large proportions. Without rapid burial, the many fish fossils found would just not exist, as the thin bones and bodies rapidly decay in days or a few weeks at best. Another good question to ask is just how do you slowly bury a giant sauropod like Brachiosaurus? What the diagrams ought to show is a violent catastrophe rapidly burying the one fish along with others near it being buried as well. Multiple layers ought to be shown piling up on top in a very short time. Finally, the fossilized fish ought to be shown with many others fossilized with it. What is also true with the book's description of the fossilization process is that problematic fossils are conveniently ignored.

*What are cases of so-called problematic fossils?*

Fossil problems include fossil graveyards. These indicate a violent and sudden death for a large number of animals all at once. These groupings of fossils have led to some strange interpretations from paleontologists. Once there were some young adult T-Rex footprints found in a group together. The conclusion was that these carnivores hunted in packs like modern wolves do.[5] The ferocious beasts may have "stuck together as a pack to increase their chances of bringing down prey and individually surviving," said study co-author Richard McCrea, a curator at the Peace Region Palaeontology Center in Canada.[6] Now isn't that something. These dinosaurs lived together, hunted together, and they probably even died together.

---

5. Ghose, "New T-Rex."
6. Ghose, "New T-Rex."

Here is what a paleontologist who found bones buried together concluded. Dr. Philip Currie, of the University of Alberta, said that evidence from ninety skeletons of Tarbosaurus Bataar—a cousin of the Tyrannosaurus Rex—suggested strongly that about half a dozen of the dinosaurs were part of a social group that died together. He said Tyrannosaurids' hunting technique may have involved juveniles chasing and catching prey, with fully grown adults taking over and delivering the fatal bites.[7] That is quite a bond to be shared among animals.

Another problematic fossil type would be polystrate tree trunks, which are buried vertically through many sedimentary layers. Google polystrate tree and see that these are partial tree trunks in a vertical position and there are many rock layers present. So how does a partial tree trunk remain upright and not decay for millions of years? How does Evolution Wrong stay upheld with so much that speaks against it? Evolution says that living things change over time. What do we do with creatures that do not change?

On page 416 of Miller and Levine's *Biology* (2008), we find a scorpion preserved in amber. Here is a fossil showing virtually no change in millions of years. The gingko leaf is another example of no changes at all. Find gingko leaf fossil images and compare them to today's leaves. The real surprise is when evolutionists say that a certain creature has been extinct basically forever, and lo and behold we find them still alive today. The *Ceolacanthus* lobe-finned fish was only known by fossils until one was caught off the coast of Madagascar. How do they know it is a *Ceolacanthus*? It looks just like the fossil forms—no change for hundreds of millions of years. Who knows which previously extinct creature will again be found?

Would a living trilobite change their minds? Definitely not. Sad to say, but evolution is such a plastic theory that it is amenable to almost anything except creation. Evolution is gradual and slow, except when it is fast. All life changes over time, except when it doesn't. The environment puts selection pressures to change the populations, which is why the same environment, like tidal waters near the shore, produces hundreds of different living things, both plant and animal, from the exact same set of pressures. Such variety from the *same* environment is not credible at all.

*Is there another type of problem associated with fossils that evolutionists often have?*

---

7. Collins, "Tyrannosaurus Rex."

## Fossils as Evidence for Which Side?

The issue discussed here has a tremendous impact on the notion that evolution has occurred. If animals and plants change over time, then the change has to start somewhere. That somewhere is called an ancestor. The major problem is that there are very few fully accepted ancestors for the major groups of living things. That this is a problem is well documented here.

This specific problem with the fossil record is nothing new. Darwin even admitted to it in *The Origin of Species*, though it was more of an apology than a defense of his theory. Dr. Andrew Knoll, Professor of Biology at Harvard University said,

> Darwin devotes two chapters of *The Origin [of Species]* to the fossil record. And you might think that's because Darwin, like most of his intellectual descendants, would have seen the fossil record as the confirmation of his theory. That you could really see, directly document, the evolution of life from the Cambrian to the present. But, in fact, when you read *The Origin [of Species]*, it turns out that Darwin's two chapters are a carefully worded apology in which he argues that natural selection is correct despite the fact that the fossils don't particularly support it.[8]

Let us explore the time between some major events in evolutionary history. Let us see if the facts and fossils have turned out to vindicate Darwin or not. The first level would be the stage in between the soft-bodied early life forms that were the ancestors of shelled life and other invertebrates. Evolutionists have often stated that this level or time period would not leave any fossils because these creatures have no hard parts and therefore would be extremely difficult, if not impossible, to fossilize, as illustrated by this quote: "Zoologists have debated the question of vertebrate origins. It has been very difficult to reconstruct lines of descent because the earliest protochordates were in all probability soft-bodied creatures that stood little chance of being preserved as fossils even under the most ideal conditions."[9] This era of time is known today as the Edicaran.

Today's fossil finds totally refute this idea. Soft-bodied organisms do leave fossils indeed. One can find worms, leaves, flowers, fish eggs, jellyfish, and even animal embryos.[10] What is even more striking is that the Edicaran life forms do not bear much resemblance at all to the ones in the following era—the Cambrian. Even more striking would be the finding of fossil sea

---

8. *Evolution*, DVD.
9. Hickman, et al, *Integrated Principles*, 485.
10. Werner, *Evolution*, 78–79.

pens or soft corals that appear exactly as today's forms do.[11] So if evolution claims that life changes through time, then why haven't these changed at all?

*Are there fossil invertebrates that are very common, yet have no ancestors?*

One of the best examples illustrating this problem would be the trilobite. Trilobites were animals related to the horseshoe crabs or other crustacean forms. They pose a special problem for evolution, as they appear fully formed suddenly, without a clear ancestral history to be found. They also possessed very complex eyes for being such "primitive" creatures. With over one hundred thousand trilobite fossils in museums and personal collections, one would think that their evolutionary development would have been worked out.

Dr. Carl Werner states the problem very well in his book, *Evolution: The Grand Experiment, Volume 1.* "If evolution occurred and if the fossil record is representative of the past, then the animals that evolved into trilobites should have been discovered by now. Despite finding hundreds of thousands of unrelated soft-bodied fossils, not one soft-bodied fossil has been declared the uncontested ancestor of trilobites."[12]

Dr. Knoll of Harvard adds, "What bothered Darwin about the fossil record more than anything else was the pattern of paleontology we have been talking about . . . the oldest fossils you see are both diverse and complex, [such as] fabulously complicated things like trilobites."[13]

This last quote gives us an opportunity to understand just how complex trilobites were focusing our attention on their eyes. Dr. Kurt Wise, the director of the Creation Research Center at Truett-McConnell College, relates a true story about the meeting of physicist Riccardo Levi-Setti and paleontologist Euan Clarkson in the early 1970s. Setti just happened to be at a talk given by Clarkson about trilobites. Setti was in attendance as trilobites had been his love since childhood. Clarkson was sharing something amazing. While studying the trilobite Order Phacopida, he noticed the eyes of these creatures had lenses unlike any other trilobites. The lenses were of just the right shape to get rid of spherical aberration, a special type of distortion that causes the edges of the image seen to be fuzzy.

---

11. Werner, *Evolution,* 93.
12. Werner, *Evolution,* 88.
13. *Evolution,* DVD.

## Fossils as Evidence for Which Side?

Now these special shapes had been discovered by human lens makers René Descartes in 1637 and by Christian Huygens in 1690.[14] Clarkson already knew that *Crozonaspis* trilobites had lenses shaped like a Descartes lens while the *Dalmanitina* type possessed the Huygens shaped lenses. This one he spoke of was unheard of in that it had both types of lens shapes in its eyes.[15]

The two scientists began a remarkable study of trilobite eyes, especially the "schizochroal" (literally "split surface") eyes. They were well suited for low light conditions, which is where trilobites apparently lived sometimes. While many compound eyes collect light mainly in the center of the field of view, these collected light from the entire field of view. They also had 3-D vision like we do, as their eyes were set more to the top of their head. Dr. Wise concludes, "Such an incredible optical system is known in no other organism, living or dead. As evidence of God's special design, truly the trilobite eyes have it."[16]

*Is there any other issue with fossils that is problematic with evolutionists?*

There is indeed. What is so interesting is that the so-called missing links are usually better interpreted as belonging to some already existing group of animals or even worse, a fake is touted as a real animal supposedly showing a stage of evolution. There is no better candidate than the famous feathered dinosaur fiasco *Archaeoraptor*. The find was given a ten-page spread in the November 1999 issue of *National Geographic* magazine. There is little wonder as to why this particular fossil caused such a stir. It was a blend of bird and dinosaur. What the magazine forgot to do was heed some sage advice. When something appears too good to be true, it usually is.

When the fossil was examined, it had remarkable features, such as feathers and wings similar to birds, but other features more reptilian, such as no feathers on the tail. *National Geographic* said, "[*Archaeoraptor liaoningensis*] is perhaps the best evidence since *Archaeopteryx* that birds did, in fact, evolve from carnivorous dinosaurs."[17]

Dr. Timothy Rowe, who had completed work on a previous "feathered dinosaur" fossil from China, which was named *Confuciusornis*, was asked by the magazine to examine this new fossil. Having experience with the

---

14. Wise, "Trilobite Eyes."
15. Wise, "Trilobite Eyes—Ultimate Optics."
16. Wise, "Trilobite Eyes—Ultimate Optics."
17. Sloan, "Feathered Dinosaurs," 100.

previous fossil, which had been altered to look as if it were a solid fossil slab, he was going to CAT scan this new one to see if it, too, had been altered. To his shock, it was not only not an entire slab, but it was a composite of several animal fossils put together.

Dr. Rowe, professor of biology and geology at the University of Texas and the director of the Vertebrate Paleontology Laboratory of the Texas Memorial Museum, states, "It was built from a new species of Cretaceous bird and a new species of a *Dromaeosaur* . . . We could find no verifiable fit between the tail, the most spectacular part of this specimen . . . and any other parts of the [fossil] block."[18] Dr. Rowe faithfully reported all of these finding to the magazine. To his amazement, *National Geographic* announced in a news conference that this fossil was indeed a genuine missing link.

But to their credit, one could ask, didn't the magazine announce a retraction for their mistake? Technically, yes they did. But it does not really merit a retraction status to some. First, it was hidden away in the forum section of the magazine, a part that not many people read. Second, it was only two sentences. Finally, it differed with the account given by Dr. Rowe as to the timeline of events. The retraction seems to say that *National Geographic* was not made aware of any problems with the fossil until March, months after the publication had been out.[19] Yet Dr. Rowe told them it was problematic three months prior to publication. Events like this might now raise doubts about any other supposed missing links or fossil finds. One must never forget that with all science there needs to be a healthy dose of skepticism. Money speaks very loudly when dealing with the fact that the emperor has no clothes.

*Are there any other significant problems with fossils?*

One last one is the issue of circular reasoning. Recall this was discussed and illustrated with a diagram in chapter 7 previously. The reason that circular reasoning is wrong is because it is like trying to successfully pull yourself up by yourself. One cannot use what he is trying to prove as a requirement to prove the first item as being true. Evolutionists frequently are guilty of this, especially when dating or working with fossils. Here is an example of two circular reasonings in one instance.

Brian Thomas, of the Institute for Creation Research, wrote about an eleven-year project at the Montana Hell Creek formation. Paleontologist

---

18. *Evolution*, DVD.
19. Werner, *Evolution*, 177.

## Fossils as Evidence for Which Side?

John Horner and three other scientists dug up and studied many *Triceratops* fossils. The scientists reported their findings in the *Proceedings of the National Academy of Sciences (PNAS)*. A quote typical of what one finds in fossil journals is given here. John Horner and his team state, "The combination of a stratigraphically controlled robust sample from the entire ☒90-m-thick HCF and identification of ontogenetic [maturation] stages makes *Triceratops* a model organism for testing hypotheses proposed for the modes of dinosaur evolution."[20] Obviously Horner and his colleagues are solid evolutionists and would not entertain anything remotely considered creation-based reasoning in the study. The study concludes that *Triceratops* evolved. This is circular reasoning, as they are using the *Triceratops* fossils to prove evolution and the "fact" of evolution to prove that *Triceratops* evolved. See the circle?

A second time circular reasoning is employed is in the cladistics analysis of these fossils. The computer programs used generated several evolutionary lineages for the fossils, some of them contradictive in nature.[21] So does this then prove that these animals evolved? How could this be construed to be proof? All the computer programs can do is spit out evolutionary plans for any data put into it. The programs are all biased toward evolution, and therefore subjective. Is not science supposed to be objective?

This chapter concludes with a quote from a professional journal. Douglas H. Erwin, Department of Paleobiology at the U.S. Museum of Natural History, said: "Resolving many evolutionary, biostratigraphic, and paleoecologic questions requires detailed stratigraphic sampling and assumes that the stratigraphic order of fossils bears some relationship to their chronological order."[22]

Now understand what he is saying. He is telling scientists that there are problems with evolution: the layers of life (biostratigraphic) and ancient environments (paleoecologic) and that to solve these issues all need to do a thorough job in getting a very good sampling of the site in question. No real problem. We all ought to be thorough in our work. The telling part is that all must *assume* that the fossils' layering bears only *some* relationship to the time order. What? Only *some*? We can thank him for his honesty. Why can't we have this truth in our textbooks?

---

20. Scannella, et al., "Evolutionary Trends," 10245–250.
21. Thomas, "Circular Arguments Punch Holes in Triceratops Study."
22. Erwin, "News Brief," 32.

If fossils do not provide evidence for evolution, then what do they point to? There are two main uses for fossils. First, they do an excellent job of showing what animals and plants looked like. We would be able to recognize a triceratops, were we to see a living one, because the fossil shows us that they had three horns on their faces and a bony frill guarding their necks. A second use for them is to join with geologic facts and tell us that there was once a worldwide devastation that not only wiped out most of life on earth, but buried it as well. What fossils do not do is show how ancient life became us.

One could ask himself two important questions: What else can cause strata layers to form worldwide? Why else are there erosion evidences missing from between sedimentary layers after supposed millions of years waiting for the next layer to be put down?

Consider that the biblical flood lasted over a year. Consider that after the initial rupturing of the earth's plates and releasing of subterranean waters (fountains of the deep Genesis 7:11) there would be tremendous amounts of water flooding over the earth. Now consider that the tides would continue to work four times per day. How much sediment could those immense tides move and carry? How many layers were laid down over a week, a month, a year? Now can you see why the flood provides a rational explanation for coal seams yards thick several times in the same place, fossil graveyards of whole collections of various animals, trees buried vertically at times, and even the finding of fossils with blood tissue in the bones or viable ink in a squid's ink sac? And I predict they would find radiocarbon in each and every soft tissue of any fossil if they'd only care to look. What are they afraid of?

There are other cases in geology that give testimony to the flood. There are massive sections of folded sedimentary rock layers. There is a video demonstrating that rocks can be deformed when under slow stress. The narration does say that it is not very much stress normally. The scientists on the video do an experiment that shows that under extreme temperature and pressure, rocks can bend without breaking. This is true, but then they set up an analogy with a limestone park bench that sags in the middle due to pressure and time. The narrator asks if a folded mountain is bent at the same rate (several inches in 150 years), then how long would it take to deform a flat sedimentary sequence a thousand feet?

If we use our common sense to investigate these assumptions, we'll remember that no one was there to see those mountains form. The flood would have tremendous Earth forces at work due to rapid subduction events. The layers were still soft like thick clay and so could be greatly bent without breaking or fracturing. Therefore, we do not need tremendous heat and pressure to bend rocks. We only need to bend them prior to their hardening.

In a closely related topic of plate tectonics, students are sometimes surprised that creationists agree with the earth science text's assertions about Pangaea and subduction zones. It must be pointed out that today's geologists assume the rate of drift was constant over time. This is unwarranted and again, no one was there to test it out. It could have been faster. Suppose the plates were drifting at a speed of ten meters per day, instead of the fastest rate of ten centimeters a year today. You think Noah would have stood after exiting the ark and said, "Wow! What a breeze!" Obviously not, however we must consider the ramifications of rapid mountain uplift. Volcanoes would have been erupting as the crust settled down to a new equilibrium of today's rates. This would set up great climactic changes.

Let's not forget that only one main Ice Age occurred after the flood. The evidences we see today, which seem to point to several ice ages, are really the growth and shrinking of the main continental glacier during the settling down period. There are excellent resources for you to continue your investigations of this phenomenon. Check out the sites listed at the end. Compare what the evolutionists say and then weigh in with the creationists.

## Chapter 10

# Textbook Evolution Evidence Explained

"It is a capital mistake to theorize before one has all the facts. Insensibly one begins to twist facts to suit theories instead of theories to suit facts."

—Sherlock Holmes, "Study in Scarlet" (Part I Chapter 3)

The Holmes quote, or more specifically the Sir Arthur Conan Doyle quote, is an excellent introduction to the main point of this chapter. I am convinced that had Darwin known back then what we know now to be true about cells and living things, the fossil record and soft tissues found in them, and even the intricacies of DNA, he would have never concocted such an idea as evolution in the macro sense. Let's now turn to biology to examine the so-called evidences that support evolution. It is well known that science teachers, especially those in public schools, are to teach evolution as the textbooks cover it. While some of the textbook evidence is scientifically true (Evolution Right), and therefore not causing any problems for biblical creationists, it is the evidence that supposedly points to *macroevolution* being true that will be examined in this chapter. It is also true that evidence for *microevolution* is then used in a "bait-and-switch" maneuver to hoodwink unsuspecting students into accepting the evolutionary history of all life on earth over billions of years. Or putting it another way, they use Evolution Right to make Evolution Wrong seem right.

*Isn't it true that evidence supports theories?*

Scientific theories are indeed supported by evidence, but no theory is ever proven true. The only thing that is an absolute truth in science is that there is no absolute truth. I realize this seems like an illogical statement, but when one thinks about it, it is perfectly true in one sense. If anyone studies the science textbooks of the past, one can find any number of facts that are now known to be not true. So science is to be always viewed with the caveat that something is true only as far as we know today.

Biology is the one science that seems to violate the above rule by demanding that one must accept everything taught about evolution as absolutely true and not at all to be questioned. If anyone thinks otherwise, he is just a yahoo, as Gould said.[1] Well, at least he was not as offensive as many evolution defenders are today. Another item to remember is that evidence is interpreted through one's worldview glasses. If one looks for evidence supporting macroevolution, he will find it. It is just as correct to say that if one looks for evidence supporting a recent creation, he will find that too.

The first item to examine, using Miller and Levine's 2012 textbook *Biology*, is biogeography. The authors say this evidence shows that the "patterns in the distribution of living and fossil species tell us how modern organisms evolved from their ancestors."[2] Darwin's finches are an example of the first use of this idea. The various finches are indeed closely related birds, yet they have differences. For example, some eat only seeds while others eat only insects. Some have bigger beaks than others. There are different patterns of colors on their feathers as well. Regardless of these differences, it is critical to remember that they are all still finches.

*What is the problem of using these birds as evidence for evolution?*

On page 473 of the Miller-Levine biology text, the Grants, a husband-wife team of evolutionary biologists at Princeton University, are mentioned as having documented that finch beaks have changed in size during a drought. Figure 10 shows how the graph appears in the text. It is drawn that way to make it seem that the beak sizes have grown over time and then stayed that way. What is not shown is that the graphed data fall back to the original

---

1. Gould, *Ever Since Darwin*, 146.
2. Miller and Levine, *Biology*, 556–57.

starting point once the rainy season hits.[3] All we have is oscillation and no true evolution in one direction.

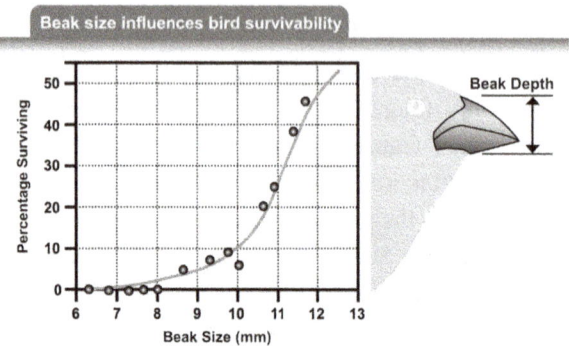

Figure 10.

The other side of this coin says that organisms can be distantly related, but similar in traits. Differences in body structure point to evolving from different ancestors, yet the similarities show how similar environments have selected similar body plans. This is why the coypu of South America and the muskrat of North America look so similar, yet have no common ancestor. True, the example comes from a 2008 version of their textbook, but still in the earlier edition, there is no explanation of what the supposed common ancestor was thought to be. And isn't evolution about evolving from common ancestors? What is this about distant ancestors? This is not exclusively evolution evidence as it also can be interpreted as limited variation of a kind. That similar environments will produce similarly adapted animals does not disprove creation.

*What about fossils as evidence for evolution?*

The next section of the text's chapter is the age of the earth and fossils, which were covered previously. Of interest is an illustration they have on pages 466–67. The evolution of whales from land animals is discussed. What is bizarre is that *Ambulocetus* (third from left; its name means walking whale) is called a whale in the text. Looking at the diagram in the book, one can see that it looks nothing like a whale. Another major flaw in the diagram is that there is no explanation of how whales with teeth suddenly became baleen whales like the blue whale shown, and then revert back to having teeth in

3. Morris, "'Beak of Finch.'"

the dolphin (*Odontocetes*). How about also explaining the huge number of structural changes as well as changes in behavior via mutations, which must be made to turn a land living dog-like animal like *Pakicetus* (second from left) into a blue whale in ten million years? It seems as if evolution waves her magic wand and things happen.

There also has been an accusation of Miller doctoring a fossil diagram for his book *Finding Darwin's God* that has the sketch of the *Ambulocetus* skeleton.[4] I have never heard of artistic license for science. At least in the 2012 edition, a photo is shown of the real fossils. Also it needs to be noted that the *Ambulocetus* drawing in the book now shows that this creature indeed did walk on all four limbs and the hind limbs are not splayed out as shown in the sketch for Miller's book *Finding Darwin's God*.

*It is not considered a continuing deception if an author corrects his earlier error. Is there any other evidence of deception?*

Were we to have two figures showing adaptive radiation from two editions of the biology textbook I have used we would see how deception is carried out in another way. Figure 11 is modeled from the 2012 textbook image which is found on p. 550 showing adaptive radiation. (Note that the earlier one used dotted lines and question marks at the bottom of the diagram to show uncertainty and a lack of transitional fossil forms.) Four years later I was quite surprised to see the lines were solid like Figure 12 shows. Were the missing links found or does artwork provide evidence of evolution?

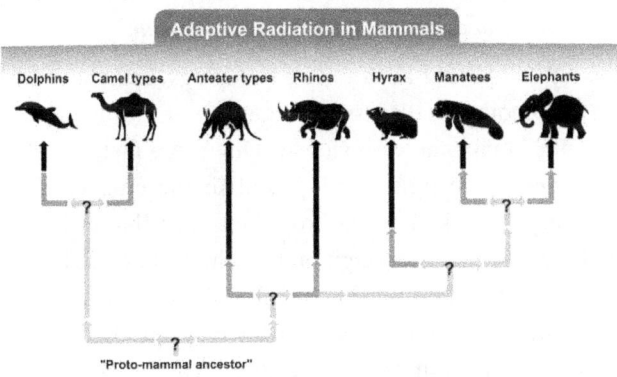

**Figure 11.**

---

4. Sarfati and Matthews, "Argument"

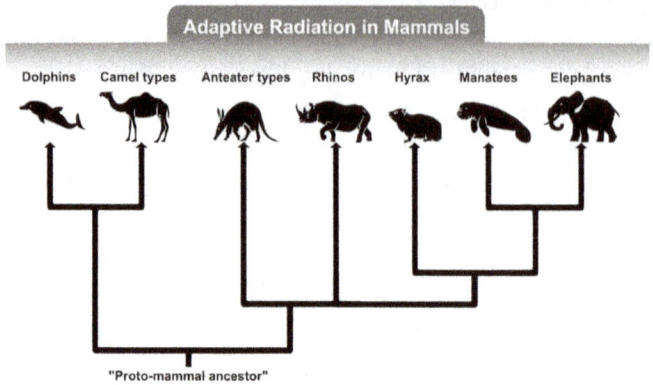

Figure 12.

Perhaps they are counting on students not questioning the teacher, following the textbook blindly. In the teacher's guide of the 2012 section for this diagram (see Fig. 11), instructors are told: "Then, guide the students in interpreting the evolutionary changes shown." In other words, don't let the students think for themselves. Teachers are to guide them like sheep. It is my wish that students would be free to question their teachers and textbooks if they feel something just isn't right. But how can they question them if both sets of diagrams are not shown?

*It is said that there is other evidence that supports evolution, namely comparing the anatomy of different creatures. Isn't this true?*

Let us look at comparing anatomy or homology and the evidence from embryology. Homology is discussed in this part of the biology text and has been dealt with in chapter 5 previously. One more fact that shoots down homology is vertebrate limb development in the embryo. Research the developmental paths taken by the frog digits and that of human fingers. The frog's "fingers" come from digit buds in the embryonic hand, while the human ones come from a solid plate in the embryonic hand.[5] Programmed cell death and other enzymes split that plate into our fingers.

How can similar features originate in such different ways? Could it be that they do not have a common ancestor? What is inconsistent with the evolutionist's take on homology has to do with convergent and analogous evolution. When two different vertebrates, such as dolphins and sharks,

---

5. Mortenson, "National Geographic."

both develop stiff pectoral fins, this is convergent evolution, but when two unrelated animals, such as insects and bats, evolve flight and have wings, then it is analogous evolution. This is how most of macroevolution or Evolution Wrong is explained. Call it a scientific name and that explains it scientifically. If you doubt it, you are accused of not being scientific at all. Evolution Wrong is a lot like the children's fairy tale of the emperor's new clothes. The emperor has clothes, but only the really intelligent can see them. Evolutionists can see evidence for evolution all around them. "Why can't you?" they will ask. You know the answer, but do they?

*They say that the environment will select the advantageous mutations. So isn't it the environment that causes evolution to move forward?*

If the environment truly does select or weed out the less fit, then what is it about a land-living environment that will select for flying? Sure enough that those insects, birds, or reptiles (pterosaurs), etc. will be more successful as they have greater range for food, can escape predators, and so on. But where in the world does all that new information come from? Not only do you have to create (or should I say "evolve") new structures never seen before, but also you need to also develop the new body support systems, like efficient respiratory organs and musculature. Then there is the aspect that these new parts need to be controlled by the brain, which means totally new behaviors and actions need to be done. That means one must mutate behavior controlled by the part of the brain controlling that part or those parts, and consider the coordination controls evolving too. Evolutionists say that evolution "created" flight four separate times (do not forget the bats), but this strains the limits of believability.

*Are you saying that the evolving of new behaviors is what really makes macroevolution impossible?*

The last area, mutating brain control over the newly acquired structure(s), is where macroevolution falls flat. Note that this topic is rarely discussed in biology texts. All the parts evolve, fine, but without the knowledge of how to use what you have just gained makes the parts all worthless. What's more is that mutated behaviors, before they are refined, makes you more susceptible to being somebody's dinner or prevents the reproduction of the species and stops evolution dead in its tracks. Imagine the first reptile—it obtains scaly skin, clawed toes, a leathery-shelled egg, internal fertilization, and a three and a half chambered heart. Okay new reptile, time to go reproduce and . . . wait . . . stop! Oh no! You have just laid your eggs in the water. You forgot to evolve the behavior to lay the eggs on land. End of the line for that "evolved" life form.

This notion of evolving behaviors brings up a very serious issue, especially if one claims to be a Christian. If behavior evolves, then why is stealing wrong? Animals do it all the time. Why is killing wrong? Animals do it all the time. Why did we evolve such things as a "law-abiding citizen" anyway? Sure, the evolutionist will spout that we have evolved altruistic genes or some such tale. But this is not supported by any rigorous testing. Animals survive just fine with no laws; why do we need them? Score one for biblical creationists having a law-giver who trumps any ideas of what humans think is fair and right. I suggest you research the evolutionary explanations for "nice behavior." Or if you happen to encounter an evolutionist, ask him to define "nice behavior" if we are just animals evolved from ape-like ancestors.

*Doesn't this idea of evolving behavior also greatly impact a Christian's notion of what belief is?*

As Luther said in his catechism, "This is most certainly true." Consider that for the evolutionist, especially the atheist ones who claim that Nature is all there is, there is no such thing as an immaterial item. Everything has a natural explanation. Therefore, there is no such thing as abstract thought. Behaviors of all types, including the "action" of believing that God or a god exists is due to the natural function of your brain. That means your faith is not based upon reality at all; instead, it is a product of your mutant fish brain. What is so ironic about this is that evolutionists usually get very upset when Christians bring God into the discussion of origins and science. Is it really the Christian's fault if his brain is not "evolved" enough as theirs is and the Christians still have a behavior that just has not been made extinct yet?

*How about vestigial organs? Doesn't evolution provide the best explanation for these?*

Vestigial organs are indeed covered next by Miller and Levine. A vestigial organ is something that had a use in the evolutionary past of an animal, but in the present is useless. These days, it is any structure that is reduced in size and has changed its function from what its ancestor used it for. The example given in the text is the small bones located where the hip would be in cetaceans. We know now that these bones help these creatures mate.[6] No mating, no cetaceans. Thus the structures have a use, and a very important one at that.

But does it make you wonder as whales were becoming whales how did hips and legs become body parts needed for successful mating? The other strange thing is that these new body parts look more different in whale[7] males and females than human male and human female hips do. Again, there needs to have been a tremendous amount of change to mutate a land animal into a cetacean. If you will study the whale evolution issue thoroughly, you will find that there are scientists who do not agree with what the textbook presents and admit that no one really knows how it happened.

What is a plus for Miller and Levine is that they have dropped the usual candidates for vestigial organs that most other textbooks still trot out. They are the appendix, your coccyx, and ear muscles. The last one is more interesting. Perhaps you know someone who can wiggle his or her ears. The muscles used to do this are supposed to be leftovers from when we were mammals that could twitch our ears to help locate a sound's direction as a horse or some dogs can do. I suppose that people who can do that are less evolved than those who can't. And you thought your Uncle Irving was just being silly when he did that.

The first two listed are hardly vestigial. Your appendix has a function in immune response when your thyroid is immature, and your coccyx helps you to stand erect and not stooped over as an ape does. Can you see why the coccyx is still called a "tailbone"? There are actually some evolutionist teachers who claim that humans used to have tails, and in fact, all people still have one while they are embryos. Tell these error-prone ones that in all cases of vertebrates with tails when they are embryos, the tail stays a tail into adulthood. Humans never had one before birth or afterward either.

Speaking of embryos, that is about the last of the evidences Miller and Levine use to convince students of the "science" behind evolution. They state that in the past, researchers noticed that embryos of vertebrates looked very much alike. Perhaps they refer to people like Ernst Haeckel. Today's facts are even more telling, they claim. Embryo tissues and organs all develop in the same order and in similar patterns. Finally there comes the errant conclusion: "Evolutionary theory offers the most logical explanation

6. Stallard, "Whales."
7. Wile, "No!"

for these similarities in patterns of development."[8] To aid in convincing the uneducated, a picture is shown. The opossum embryo is touted as being so alike with other vertebrate ones, that only an expert can tell that this is an opossum embryo.

I guess the previous is true, as I have never seen a mouse, raccoon, and opossum embryo side by side. But let us examine the history of this incorrect interpretation. Ernst Haeckel, mentioned above, was one of the first popularizers of this embryo likeness idea. In fact, Darwin considered this evidence to be excellent and referred to it in his work. Haeckel's sketches were printed in biology texts for decades, were in the first biology texts I taught from, and can probably be still found in textbooks today. Figure 13 shows you Haeckel's original sketches. Were these true and accurate sketches, then Miller and Levine could be correct in saying that embryo likenesses might point to a common ancestor.

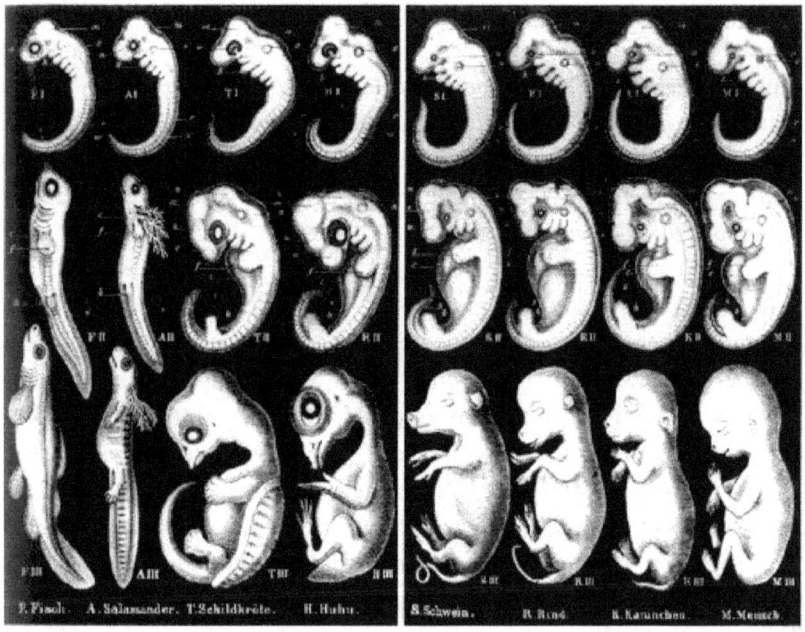

Figure 13.

To the duo's credit, this misinformation is not printed in the 2012 edition. Instead, they use an opossum embryo and say that you need to be an expert to tell what it is. This is very misleading. Study the sketches of what

8. Miller and Levine, *Biology*, 469.

vertebrate embryos really look like in Figure 14. Do you think you can tell them apart? Do I think Miller and Levine would now call you an expert? I think not, but the point has been made.

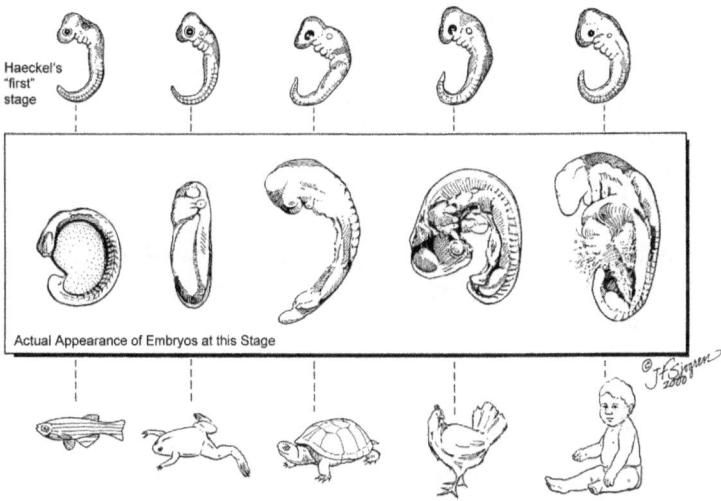

**Figure 14.**

There was a prior biology text I taught from which used Haeckel's embryos for the evolution chapter, and then in a section of reproduction, it showed very early vertebrate embryos. The students were to note how different they were from each other. I would point out to students how one was science fact (the latter) and the other was science fraud (the former). It is sad to see science's fraudulent ideas being used to prop up what I feel is a major misguided notion masquerading as science.

*Just what, then, do embryos point to?*

First of all, it is no surprise that there is some superficial resemblance among vertebrate embryos. What else can you do with a head, tail, and four limbs attached to a backbone? Consider this analogy. There are many variations of cars, but they all share the same basic characteristics: doors, windows, engine, steering wheel, and so on. These cars are products of intelligent design, just as God has intelligently designed all life. It certainly points to wisdom to have variations of a basic design incorporated into living things. No evolutionist is upset when cars all basically look the same and they all agree that these are designed; but try to have them remain calm if a student

## Is Evolution Compatible with Christianity?

holds that living things, which are vastly more complex than any car, are products of a supreme mind.

*What is this I hear about molecular evidence for evolution?*

The last evidence the authors mention is molecular in nature. The two best candidates for molecular evidence of evolution are DNA and homologous molecules like hemoglobin or cytochrome C. The idea is that molecules of closely related organisms will show a similar close relationship of sequences, and more distantly related animals will show a larger difference in those same sequences. For example, you may have heard that chimp and human DNA differ by only 2 percent of DNA bases compared. There is even supposed evidence that purportedly shows that one of our chromosomes was a fusion of two chimp chromosomes.

In the 2012 edition of the biology textbook, there are small "Analyzing Data" exercises for the students to do. There is one associated with this very topic. On page 470 is an exercise asking students to compare three strings of DNA. Each strand is shown as forty boxes and within each box is one of the base letters A, C, T, or G. The creatures supplying the DNA are a mouse, a baleen whale, and a chicken. The percentage of difference is predictable. The whale is only 10 percent different from the mouse, while the chicken is 20 percent, unlike the mouse's strand. The students are obviously forced to conclude that the mouse is closely related to the whale as the percentage of difference is much less. In cards when one stacks the deck, it is not hard to get the result one wants. This exercise is a perfect example of stacking the deck. Carefully setup the DNA strands printed in the book, and the students will reach the conclusion the authors wanted everyone to get.

An interesting evaluative question is asked of the students. They are asked if they think scientists use small snips of DNA to infer evolutionary relationships. The answer key says that the student ought to answer no. The answer continues that scientists should use longer sections of DNA. The key missing data would be to know just how long the strands are which are used for comparison—one thousand bases? Ten thousand? One million?

Here is the bad news. If Miller and Levine really were interested in critical thinking, they ought to have students analyze data from animals that are said to be more disparate in percentage, yet are closer than another member of the same order or family. There are plenty of discords to be found in genetic research. For example, read this quote: "Indeed, at 6 million years of separation, the difference in MSY gene content in chimpanzee and human is more comparable to the difference in autosomal gene content in chicken and human, at 310 million years of separation."[9] Another interesting comparison would be to look at chromosome numbers of organisms. Those closely related evolutionarily ought to have similar numbers of chromosomes, but the chromosome number seems to have nothing to do with evolution.

*Wouldn't changing the chromosome number of an organism be detrimental to that organism?*

Indeed so. Research the problems that happen if one of our chromosomes is missing or extra. There is only one condition, called Triple X, where a female has three X chromosomes, but has no obvious issues or conditions. Stranger still are the chromosome number differences within the horse kind. For example, horses have sixty-four, donkeys have sixty-two, and zebras can have thirty-two, forty-four, or forty-six.[10] Dr. Jean Lightner, a creationist who has written considerably on this topic, says this about the horse chromosomes:

> The differences in chromosome number in these different members of the horse kind show that changes have occurred in how the DNA (genetic information) is packaged. Some of the changes are a bit complex: separate chromosomes have fused together and some centromeres appear to have changed position. However, this is not evolution in the sense that anything truly new has arisen. Rather it is a repackaging of what already existed.
>
> Rearrangements in chromosomes do not appear to be the result of just chance, random processes. It is very important that such changes occur in a way the integrity of the information is maintained. In order for that to be accomplished, it would seem cellular mechanisms would be required to regulate these relatively rare events.[11]

Note the key words in the last paragraph: information, mechanisms, and regulate. These certainly do not sound like the workings of mutations at all. Rather, they speak of design, which obviously point to a designer.

More bad news would be found in homologous molecules of different animals. One can find very dissimilar molecule sequences in animals

---

9. Hughes, et al., "Chimpanzee and Human," 536–39.
10. Mitchell, "Truth from Telegraph."
11. Mitchell, "Truth from Telegraph."

that are thought to be closely related evolutionarily. Again, the authors have cherry-picked the data with as much gusto as creationists are supposed to cherry pick quotes from the evolutionists' writings. The difference is the evolutionist makes the data say things it was never meant to say. Creationists make the evolutionist say things exactly how they meant them, but the public isn't supposed to know they said that item. Which side really is interested in the truth?

*So what is the reason that you feel molecules are so closely related yet they do not point to a common ancestor?*

My personal opinion as to why the molecules are so closely alike is quite simple: we have to eat. There it is. It is just that simple. I tell my students that if you want to make DNA for your body cells, then one must eat DNA. Imagine how complicated life would be if all life were not designed to be so compatible. Humans and other living things would have very little to eat and death would soon result.

Let us summarize: the data is what it is. One can make it say just about anything. The data can support a Bible-based outlook on life and the cosmos, if one looks at it from our perspective. Can it support the Evolution Wrong conclusion? Only if one plays games with the data and hides what is troublesome. I hold that were evolution not supported by high government grants, then scientists would not be so concerned about doing a philosopher's or theologian's job and try to tell people where they came from and how everything got here. We could be applying our cerebral horsepower to real issues like cancer or viable energy sources that do not pollute as much.

# Chapter 11

# The Origins of Everything

> Sherlock Holmes: "What do you say to that, Watson?"
> Watson: "Well, it is all possible if you will grant the original monstrous supposition."
>
> —"The Adventure of Shoscombe Old Place"

In the beginning, Nature caused the first infinitely dense particle to come into existence from nothing. This particle then exploded, creating the space-time continuum, and after cooling, the first atoms of hydrogen and helium were formed. These atoms collected together into the first stars, and they eventually exploded also after creating many of the heavier elements, such as iron, in their cores. Those stellar novae threw the heavier elements out into ever-expanding space to create more protostars surrounded by gaseous nebulae. The nebulae coalesced into planets. One of these planets just happened to have the fortuitous mix of elements that caused the first organic molecules to create themselves. Eventually, a membranous envelope surrounded the now reproducing biopolymers, and cells began. Smaller living cells, the ancestors of today's prokaryotes, soon were ingested by the membrane envelopes to become more complex.

Throughout this book, I have mentioned the issues of origins. Now we can focus on the details of the origin of life as Evolution Wrong tells it. As the saying goes, the devil is in the details, and it is the details the evolutionists wish to keep under wraps. There are so many miraculous events occurring in the evolutionary origins account that it almost seems inappropriate to call atheist evolutionists atheists. Their god is Nature, and that, as stated before in chapter 4, is why this word is capitalized so often in the middle of a sentence.

*If "Nature" is like a god, shouldn't it do miracles?*

One example of Nature's miracles happened when our solar system came to be. Today's modern theory began in the 1960s. Astronomer Dr. Danny Faulkner relates particulars of this idea in a chapter of *The New Answers Book 4*. Referring to previous material describing what astronomers of the mid to late 1700s thought, he says,

> As before, the solar system supposedly formed from the collapse of a gas cloud that flattened and concentrated in its center, with the sun forming from the central condensation and the planets forming from the material in the disk. The modern theory borrows a term coined by Chamberlin, [Thomas Chamberlin, 1905] *planetesimal* (from the word "planet" and "infinitesimal"). A planetesimal is a small body amalgamated from microscopic particles. Planetesimals supposedly grew in the proto-planetary disk to form bodies large enough to begin gravitationally attracting other planetesimals to form the planets.[1]

The miracles of "Nature" begin here. She waves her magic wand and solves many problems that demolish the validity of this idea. Besides the continuing issue with angular momentum there are others. One is just what caused the tiny bits of matter to come together into chunks of rock large enough to form a gravity field? Modern astronomers suggest several less than possible solutions to the clumping problem. One is that the particles were covered with a gooey substance that stuck them together. Another is that some sort of static electricity was at work.[2] Another problem is to ask just what caused the initial gas cloud to collapse? Faulkner shares that some think a supernova shockwave could cause the nebula to contract so star building can begin. But this answer is a problem in itself. Think about it. You need a star to blow up to make a supernova, but you need a supernova to make a star. This is a typical chicken/egg quandary in which the evolutionists find themselves. It is similar to the adage that it takes money to

1. Ham, *New Answers*, 370.
2. Ham, *New Answers*, 370.

make money, but the evolutionists are printing counterfeit cash to convince you to invest in their scheme.

The planetesimal theory was explained to me while I was in a master's degree program at a local university. The professor was a superior teacher who brought concepts to an understandable level. What was even more surprising was that during the entire course, not one hint of evolution was said, planetary or otherwise. I had to wait until the course was almost over before one student asked about the origin of the solar system. The professor began his tale with these words: "Let me tell you what actually happened." At that moment I knew he was being unscientific. I waited for him to finish. When he stopped, I rose my hand, was called on, and asked, "How do you know what actually happened? Were you actually there?" It was almost like Figure 15 here.

Figure 15.

The lecture hall's reaction caught me by surprise. Most everyone burst out laughing. The professor laughed, too, but looked more like a deer in headlights. This was due to the fact that he got caught with his scientific pants down. After the laughing stopped, he responded, "Well that is what astronomers believe." Wow! He used a religious word for a "scientific" idea. That was a huge admission. The reader must know that he did not mean "believe" in a religious way, but the question asked got him to take his tale out of science fact (actually happened) to science fantasy (believe).

As far as the origin for the moon, he said that no one at NASA accepts the capture theory for the moon's origin, as it is mathematically impossible. The capture theory stated that a wandering planetoid, our moon, just happened to get too close to the earth and got captured by Earth's gravity into the orbit it has now.[3] I guess that is why evolutionists now have a Mars-sized planet hitting the earth to make the moon (see following paragraph).

*What issues are there with this origin idea that was accepted by a consensus of planetary scientists back in 1984?*[4]

According to the presently accepted theory, a Mars-sized planet struck the young earth at a glancing blow. Animations show that the materials from the object and the earth somehow coalesced into the present moon. It is just truly amazing what intelligently designed computer generated animations can do to "prove" that it all happened by chance alone. One major problem has recently been uncovered. Earth rocks contain trace amounts of titanium. The moon rocks brought back by the Apollo missions have the same ratio of titanium as Earth rocks do. Therefore, the impactor's rocks are nowhere to be found in any of the samples.[5]

Brian Thomas[6], a contributor for ICR articles, adds this from his research:

> An increasing number of computer simulations have revealed additional flaws in this planetary collision model. Science journalist Daniel Clery recently wrote in *Science*, "As a result, researchers are casting around for new explanations. At a meeting at the Royal Society in London last month—the first devoted to moon formation in 15 years—experts reviewed the evidence. They ended the meeting in an even deeper impasse than before, as several proposed solutions to the moon puzzle were found wanting."[7]

A very revealing quote comes from the same article said by David Stevenson, who helped to put the event together. He said, "It's got people thinking about the direction we need to go to find a story that makes sense."[8] So now scientists are storytellers? This is the entire problem with science concerning the issue of origins. Most of the people doing it are trying to find a Nature answer or story for a supernatural event. So will they ever get it correct? As the youth of the 90s used to say, "Not!"

3. Dejoie and Truelove, "StarChild Question."
4. Dejoie and Truelove, "StarChild Question."
5. Samec, "Lunar Formation."
6. Thomas, "Impact Theory."
7. Clery, "Impact Theory," 183–85.
8. Clery, "Impact Theory," 183–85.

# The Origins of Everything

*What about the creation of life?*

The next major step evolution must make is the creation of life from non-living inorganic molecules. Here is where the objective science of chemistry is no friend to the idea that chemicals randomly combined to make more complex ones. Keep in mind that not only do these molecules have to become more complex, but they also have to spell something genetically. Evolutionists claim that information created itself. There is a better chance that this paragraph can be put together randomly than for an average body protein to be put together randomly.

One way evolutionists explain in science textbooks how this happened is to use a device I have dubbed "magic arrows." Once you understand what these are, you will find them used quite readily by illustrators of biology texts. Now it is critical to understand that the artist is not trying to explain evolution, but it is the author of the text telling the artist what to draw.

Basically it is an arrow in between two other items, be they animals, molecules, or whatever. Now the caption will say something like the diagram "shows" how this event happened (where one item changes into the other item). Apply this thinking to any magic trick you see. When it is completed, what is the typical response of anyone seeing it? "How did you do that?" Of course, once the trick is explained, then all can see it is not magic at all, but really an illusion. Anyone who just had a light bulb go off in his or her mind, you are learning to see what most of macroevolution is—an illusion. It is an attempt to have you see something that is not really there.

On one side of the arrow we might have a jumble of molecules, and the other side has biopolymers, or you can see a fish on the left and the other side has a salamander-like amphibian. Ask the evolutionist, "How did nature do that?" The typical response is that the fish "evolved" or "developed" or "solved the problem of _____," and you are just supposed to swallow that. If you are brave and know a thing or two about basic genetics, you might ask more probing questions like how do mutations create something, like a cardio-pulmonary system that was not there in the fish. But be prepared. When one of my student's friends questioned her public school science teacher about evolution, the teacher yelled at her. Remind your friends to be courageous and when yelled at to report this unprofessional conduct to your guidance counselor and the principal. Include signed statements of student witnesses, which will be evidence supporting your case. Asking questions must be protected. How else can true learning happen?

## Is Evolution Compatible with Christianity?

This is the cue for Toto to run to the curtain and show you the wizards of evolution are not who they portray themselves to be. Any scientist ought to be able to reproduce in the lab what is taught in a book. Or at least you ought to be able to see it for yourself, as in mitosis slides that are actual cells in the process of dividing. But with evolution, one must "pay no attention to the man behind the curtain." See for yourself that there is nothing they can show to prove that what they say is true. This is why they need artwork to prove evolution instead of lab experiments. As a short sidestep here, see the diagrams below from one of my PowerPoint presentations.

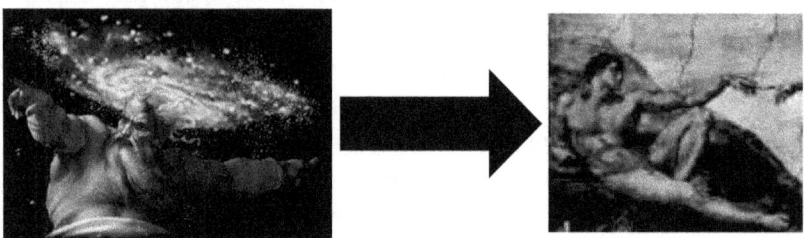

**Figure 16.**

Note that I have a magic arrow between "God" and "Adam." This now fully explains how God created Adam as stated in Genesis. Surely this is scientifically valid and thus is not religious in any sense as I have just used what they draw. Artwork is proving the point. If they can use art to make their point, so can anyone.

*Let's assume that life did create itself. The next step would be to make cells then, right?*

Continuing further along with the evolution explanation of how life got to be where it is today from the Big Bang, which has problems of its own if you would care to read Dr. Jason Lisle's works and essays, we come now to how cells became more complex by the process of organelle acquisition. Let's read how it happened from Miller and Levine.

> Researchers hypothesize that about 2 billion years ago, some ancient prokaryotes began evolving internal cell membranes. These prokaryotes were the ancestors of eukaryotic organisms. Then, according to endosymbiotic theory, prokaryotic cells entered those ancestral eukaryotes. These intruders didn't infect their hosts, as parasites would have done, and the host cells didn't digest them, as they would have digested prey. Instead the small

prokaryotes began living inside the larger cells, as shown in [the figure below].⁹

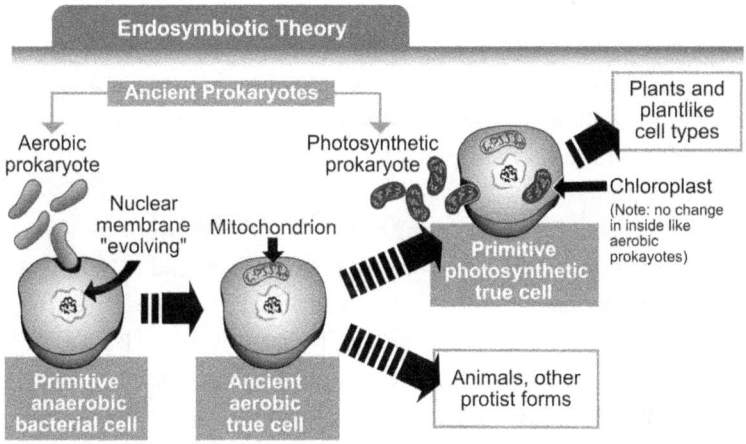

**Figure 17.**

Okay, see Figure 17, which is modeled after the text's image. Note the magic arrows. What is odd is that in the 2008 edition, at the start of the section "Endosymbiotic Theory," Miller-Levine say this, "Then, something radical seems to have happened." The radical things are that the invaders were not digested and the invaders did not infect their hosts. Why did they drop this sentence in the 2012 edition for the same section? Is it no longer radical? It must be common practice for bacteria to enter host cells and not make them sick. It must be an everyday thing for eukaryotic cells to ingest bacteria and not have them for lunch. Here we have nature doing her miraculous work again. Poof, we now have a symbiont! Luckily these went on to become us or we wouldn't be here to read this book and have this debate on how we got here. Funny how evolutionists do not like God or miracles, as they feel he breaks rules. Not surprisingly it is the evolutionists' god Nature who routinely does miracles breaking scientific rules to evolve things, and funny those evolutionists say it is scientific for Nature to break rules. "Does evolution even have rules?" makes for a great question begging to be asked.

*What even led to this idea of endosymbiosis?*

---

9. Miller and Levine, *Biology*, 556.

## Is Evolution Compatible with Christianity?

To be fair, there is some evidence that led Dr. Lynn Margulis, the creator of the endosymbiont theory, to suggest this amazing happenstance. It seems that mitochondria and chloroplasts have DNA similar to that of bacteria. Additionally, both have ribosomes similar to those found in bacteria. Finally, both reproduce by binary fission like bacteria, so that when cells reproduce there will be enough new mitochondria/chloroplasts for the daughter cells. Thus they say that these evidences provide strong support that mitochondria and chloroplasts are the descendants of the ancestral invader symbionts.

Or they could provide strong support that God uses what works, just like current designers do. A good analogy for this is found in NASA's Apollo Program. Parts were designed first for some high tech need, like computer chips for the Apollo command module. Today those chips led the pathway for current laptops using that same design, but altered for our use.

Another issue with endosymbiosis is that the insides of the organelles in question are much more structured than those of bacteria. Consult any biology text or website showing the parts of any bacterial cell compared to that of any eukaryotic cell. Also carefully note that it is an insult of sorts to call bacterial cells simple. If they are so simple, why can't we make one?

Finally, how much of a staggering coincidence is it that mitochondria and chloroplasts just happened to give off what the ancestral cell host needed, and the host cell just happened to provide the nutrients needed by the invader. Add to this the fact that the chloroplasts and mitochondria both produce waste products that the other uses to make what the other organelle needs. This is just too good to be true; hence you need to be aware of the "just so" story often used by evolutionists. I call it the Goldilocks Syndrome. Everything is "just right."

*Hold on . . . what is meant by a "just so" story? Can you give an example?*

Consider an analogy found in an Answers in Genesis article. The topic covered in the article was the fact that evolutionists have discovered that comb jellies (creatures related to jellyfish) diverged and became more complex way before the less complex sponges did.[10] The AIG article then provides the analogy of using circuit board technology.

A somewhat loose analogy might be this: suppose someone took it on faith that all computer circuit boards were simply modifications of modifications of modifications (and so on) of the original circuit board—that

---

10. National Science Foundation, "First Animal."

factory mistakes here or there had resulted in the diversity of computer circuit boards we use today (assuming, of course, at least some of these mistakes weren't detrimental to the boards). Now, suppose this person, after devising a complicated way to standardize and measure the exact difference between any two working circuit boards, determined that what looks like a modern, high-tech circuit board was actually older and more original than a comparably low-tech, simplistic circuit board. He would at that point have only three options:

> 1. "This doesn't make any sense; my system for measuring and comparing differences—and concluding which board is the oldest—must be flawed in some way."
>
> 2. "This completely disputes the idea of descent with increasingly higher-tech modification for circuit boards (since we now have many more high-tech boards); maybe something else was responsible for the origin of all these circuit boards."
>
> 3. "I know that these circuit boards all descended from one original board design, and that means my system of measuring differences must be correct. So I'm just going to have to come up with some explanation for why such an unexpected result occurred."[11]

In the last option, our hypothetical believer in circuit board descent would employ a "rescuing device": a "just-so" story that is held by faith but that allows him to hang on to his contradicting beliefs.[12]

*How do Miller and Levine explain the evolution of sexual reproduction?*

The final part of Miller and Levine's biology textbook's section on cellular origins covers sexual reproduction and multicellular life, in that order. Sexual reproduction is the union of two gametes: sperm and egg. Miller and Levine explain that bacteria reproduce daughter cells with genomes duplicate of the mother cell. This limits evolution to just the mutations of the DNA. They then state that eukaryotes reproduce sexually. This increases the variability of the offspring so the daughter cells are different from the parents.

Now I have never heard of sexual reproduction among single-celled protists. The only process coming even close is conjugation done by paramecium. Nowhere can one find this labeled as sexual reproduction. It has been called "primitive sexual reproduction" in other textbooks, but not true

11. First Creature on Earth.
12. "First Creature."

sexual reproduction. Plus, paramecia are today's cells. The ancient protists had to have been much less evolved, so how does something as complex as sex "just happen"? And don't you need it to happen differently to each sex so you evolve a male and a female? But what are the odds of them mutating at the exact same time? If evolutionists played the lottery, they ought to be billionaires by now, since chance is a god who makes miracles happen routinely.

The 2008 edition had a neat analogy about sexual reproduction. On page 428, Miller and Levine likened the genetic recombination to the shuffling of a deck of cards. Think of how many different hands one can get from one deck is the analogy. This analogy is good when just considering the reshuffling of genes done by crossing over coupled with independent assortment. But evolution requires more than just different hands. You also must evolve different games that would represent the different types of living this on earth. The evolutionist might ask if you mean like war, solitaire, rummy, bridge, euchre, etc.? You ought to respond no, I mean shuffling a standard deck will give you an Uno deck, with the wild draw four cards, reverse, and others, *and* from shuffling we will eventually get poker chips to evolve betting games, *and* from shuffling a deck of cards you will evolve a computer to play card games on. If one is impossible, so is the other. Shuffling genes will not evolve new animals with structures and parts never had before. As stated earlier in this book, fruit fly studies destroy this notion that mutations are how novel features become part of an animal or plant. No geneticist can get fruit flies to mutate one single thing that is a new beneficial structure. We can only warp or mutate what is already there.

# Chapter 12

# Stephen Meyer Weighs in and You Should Too

*"Now, I make it a point of never having any prejudices, and of following docilely wherever fact may lead me."*

—"The Reigate Puzzle"

A MAJOR BOOK CONCERNING intelligent design is Stephen Meyer's *The Signature in the Cell*. Besides covering some important ideas and evidences that support Meyer's thesis, the concept of intelligent design will be discussed here. Meyer's book does a thorough job of exposing the weaknesses and flaws of the various attempts by evolutionists to explain the origins of life and the information needed to cause that life to happen. He covers the evolutionary ideas of DNA being first, RNA being first, and protein being first hypotheses. He also does a good job explaining that information only comes from a mind in this present time. This chapter will conclude with the tactics of the new atheists and how to counter them.

*In discussing intelligent design, what would be the most intelligent questions to ask?*

The key questions to ask are these: Is information an intelligently designed item? If it is not, then what causes it to be? Other critical questions are the following: (1) Are the instructions for protein synthesis akin to instructions on how to build a car? (2) Are instructions a form of a message from a source to builder? (3) Is information a property inherent in matter? (4) Are there any natural means for creating information, instructions, or messages outside of an intelligent mind? If such natural means exist, what are they?

I recall asking some evolutionists this sort of thing on a blog I was invited to join. I would caution anyone in joining blogs hostile to Christianity in general or creation specifically. Most of the members had language that resembled a junior high locker room. People trying to make themselves sound intelligent with four-letter words does not convince anyone of their intellectual stature, except to impress other small minds. They also like to gang up on you. Finally, they have the most obtuse ideas. I asked them to show me a message with no intelligence behind it. One person responded, seriously suggesting that the earth talks to us via tremors, but we just do not know the language yet. I told him to get a pet rock and maybe it would obey his commands to stay, if he was persistent.

*How does Meyer begin dealing with this topic of information?*

Meyer sets the stage for the discussion with object lessons from his own university teaching. Using Scrabble letters, he asked his students to draw a letter from the bag, write the letter on the board, return the letter to the bag, and have the next student repeat the process. This continued until there would be a set of letters that would spell a short word like "run." Students were ecstatic. But the goal was to create a sentence of seven or so words, or at least a word of ten or so letters in length. Students soon saw that chance just could not do it.

*Haven't other experiments shown that information can be randomly created?*

Some would ask if those are examples of chance creating long sentences. It would be amazing evidence indeed, except evolutionists cheat—again. Dawkins and others got a computer to create a line from Shakespeare: "Methinks it is like a weasel." How he and his cohorts cheated was that they programmed the computer to select for any letters that randomly fell into the proper slot, thus the computer kept any letter that was correct.[1] Nature cannot keep correctly placed amino acids in a mostly incorrect polypeptide. The entire chain must be correct in order for nature to select it. Plus,

---

1. Meyer, *Signature in Cell*, 281.

they had a target sentence or goal they aimed for. A good question to ask any evolutionist is that if evolution has no goal and is blind, how could it know to "aim for" that target sentence? Thus Dawkins et al. did not reflect the nature of evolution, much less the nature of nature. Finally, is it not the height of arrogance to use an intelligently designed computer and software programming to prove that an intelligent designer is not needed?

Evolutionists could point to Avida, but it is full of cheating also. It is a computer created "world" filled with computer designed organisms that are full of preexisting information, like the ability to replicate. To be fair, this computer program was not looking for an intended target. It was more ambitious. It generated an "Avida World" populated by Avida organisms. How this program mimics evolution is that upon copying an organism's "DNA," the Avida World makes random changes to the organism's programing by switching, deleting, or inserting new commands. Next, an evaluation algorithm (an information rich set of rules) searches for organisms that can do more tasks.[2] This is the part that is supposed to be like Nature "choosing" the fittest to survive. Not lost on me is the utter irony that this algorithm mimicking "random" Nature is very intelligently designed. So these "organisms" are selected by something that is looking for something . . . like fit, digital organisms.

The last issue is the most critical, in my opinion. Again, we have an information-rich "world" created by intelligent agents that supposedly mimic how Nature works with no intelligence at all. Add to this that the complexity of the digital organisms, or the "world" in which they exist, is nowhere near the complexity of even a single real biological protein or enzyme. Nice try, evolutionist.

*How does one go about choosing a proper idea or theory of origins?*

The first criterion is that it is logical. For example, one cannot have fire and ice exist in the same place at the same time. Next, one must make certain it does not violate any science laws. It is here that the evolutionist will object strenuously because Christians allow miracles to occur. They state that miracles violate natural laws. Sorry to disappoint you materialists, but miracles obey a higher law. Consider fire trucks racing to a call. The law says that all vehicles will stop at a red light. But when a fire truck goes through a red light with sirens blaring, they are not disobeying a traffic law, but obeying a higher law that says that they can. Consider how Jesus,

---

2. Meyer, *Signature in Cell*, 287.

after the resurrection, had the ability to walk through walls. Disobeying natural law? No, he was then obeying the higher law of how his glorified body operated.

The last major item an origin theory needs is to have the idea fit observed facts seen in the lab or in nature. Do we see anything in nature, outside of living things, using anything like a code or instructions? Of course we do not. All instructions or codes have a mind or an intelligent agent behind them. Here now is where we can explain what specified complexity, a key component of intelligent design, is.

Examine the following three lines:
A. lkjfwq;omxc,vnmn
B. lmnoplmnoplmnoplmnop
C. Jesus loves me and he died for my sins.

Line A is gibberish and is what random chance will produce. This is akin to what the Urey-Miller apparatus produced. It did produce amino acids, yes. But it also created biological gobbledygook. Line B is highly ordered but conveys no information. An example of this is an ice or quartz crystal. Nice pattern, but it doesn't convey any message. Line C is specified complexity. We can tell this because any change, no matter how small, will not just simply change the meaning, but will most likely cause the item to lose meaning altogether. This is how you can know that DNA did not put itself together.

*Why does DNA or other biological facts of any cell cause problems for evolutionists?*

Evolutionists are hard pressed to describe something like DNA without using words that are reserved for designed items. You have read several in this book. DNA has been called the code or instructions for protein assembly. Others have described it as being an information rich molecule. Interesting to note that there are articles where evolutionists ask their colleagues not to used "design-friendly" language anymore. The problem is that we have not created any design-free implication words to describe designed items.[3]

Even more incredible is when scientists use design-friendly terms and then in the same breath or sentence deny that they are implying a designer. Dr. Jerry Bergman reports such a case with this gem: "In the next example, evolutionists Stephen Hetz and Timothy Bradley, both of the University of California, Irvine, stated that the insect respiratory system was '*designed* to

---

3. Bergman, "Scientists Urge Censorship."

function most efficiently at high levels of O2 consumption' adding that they 'did not intend to imply that the insect's tracheal system is the result of the work of a designer.'"[4] They added that the term *design* is a "shorthand for an awful lot of ideas, such as that the system has been shaped by selection pressures to have a certain functional consequence."[5]

The language used by evolutionists is very crucial. I have had my students read the text and do a word count in the evolution chapter. They were to find "language of doubt words," such as *probably, possibly, should have, could have, might have*, the "been" cousins of the last three, and *speculate. Speculate* is an interesting word. All it means is "guess." It is important to stress here that there is nothing wrong with scientists guessing. What is wrong is when they call a guess a fact. It should be noted that my students counted those doubt words and found around fourteen instances. Next, I had them read the same number of pages on the chapter of mitosis or cell respiration. The count was around two or three. The main idea behind this exercise was to show that according to evolutionists, evolution (Evolution Wrong) is a fact—we just have no idea what is going on or how it happens or why it occurs. With more recent textbook editions though, I find that this activity is almost useless. It seems that evolution authors are becoming less honest and simply dropping those words. Now students are mistakenly led to believe that evolution's ideas and guesses are solid, lab-tested facts.

*I have heard that intelligent design is not anything like a science. Isn't it really just a religious idea dressed up or masquerading as science?*

One of the biggest complaints against intelligent design is that it is religion masquerading as science. In the book *Creationist's Trojan Horse*, the authors make a case for the motives of some in the intelligent design movement. But does this nullify what arguments the intelligent design side makes? What about all those evolutionists who are openly hostile to Christians and who push materialism? Don't they have a motive or an agenda that is "religious" as well? Of course they do. As an example, here is an actual e-mail I received from an evolutionist teacher.

> You are not preparing your students to deal with people like me at all. You are, in fact, denying them the tools they need to interpret and understand the world around them, and are insuring that they enter that world unable to make rational arguments to support

4. Flores, "Journals," 12.
5. Quoted in Flores, "Journals," 12.

> their own ideas (if they ever even have their own ideas.) You are not acting the part of a Biology teacher, but that of a religious preacher indoctrinating children into a colorful, entertaining, and comforting world of self-deception.
>
> If you want them to really be able to hold their own against evil people like me, you'd better start teaching them how Darwin and a host of others arrived at their conclusions, and precisely what those conclusions are. If you don't, I assure you that if someone like me gets a hold of them they'll crack like hot glass in ice water as soon as they hear the truth presented as evidence rather than silly parables and aphorisms. I've seen it happen many times.

The main reason that this teacher has seen young Christians "crack" so many times is that most churches do not train their youth in apologetics. Were they to know what any well-read Christian trained in defending his faith ought to know, they would be "crack" resistant. Thus they shatter because of a lack of training, not due to the "science facts" of evolution. Plus, note the tone of this teacher. He would be intimidating to any young person. He may well ridicule them or make them feel stupid and not "scientific" if they stand by their faith. Such tactics do not belong in the science classroom, and just might be more common than believed.

Additionally, notice that this teacher who emailed me feels it is his job to cause students to give up or renounce their faith. I have no idea what state standard requires biology teachers to do this. Next is his error that creationist science teachers, like I am, do not teach Darwin's and others' conclusions and how they reached those conclusions. I do teach the evidence, but the students interpret that evidence with their Christian worldview intact. It is true that he does the same thing, but at least I am honest enough to tell the students about worldviews and how they color evidence to make it say what the person wants. Sure, I am guilty of this as well, but we Christians have God on our side. Who does he have? Nature?

It is ironic that he talks about truth when in previous e-mails he complained that I kept harping on the issue of truth. It annoyed him. I would ask him the question Pilate asked Jesus, "What is truth?" Is truth just a notion from our mutant fish genes? If so, then there is no such thing as truth. Now consider seriously how this impacts Christianity in a most serious way. Jesus said, "I am the Way, the Truth, and the Life" (John 14:6). If there are no absolutes, then what does each of those critical words mean? In fact, how can we even read this book or anything unless meanings mean something and do not change? Makes one's head spin. By the same token, consider that nothing the other side says makes sense either as no one knows what anyone says or means.

Another different type of question arises now. Did this teacher sound like someone with an agenda to you? Why not present the facts for both sides and let the kids make up their own minds? Or one can check out P. Z. Meyers, Richard Dawkins, or any of the other big name evolutionists. You will find them involved in name-calling, belittling, anger, mocking, and so on. This is hardly the high road to take if one has a solid case to stand on. On no other scientific topic does one see this. To add one more nail in the coffin, I received a post card from PBS about their new (at that time) *Evolution* series. One of the reviewers after watching the series said it was "a near religious experience." What exactly does an atheistic or naturalistic series reviewer mean by this? I have always heard these evolution defenders say that science has nothing to do with religion.

The reason that intelligent design (ID) makes sense is that it is logical. Because it is logical and flows from the facts and data gathered, most evolutionists create a "straw man" to battle and defeat. Here is one of the straw men the evolutionists say ID is. Premise one: material causes cannot produce or explain specified information. Conclusion: therefore an intelligent cause must have produced biological specified information. This is called arguing from ignorance. It is what most of my students say about life or some specific item in biology. "This is so complex that God had to do it." Perhaps if I tell them that when they use this form of logic, they are sounding ignorant, then they might stop using it.

Meyer points out that the ID logic goes like this instead. Premise one: despite a thorough search, no material causes have been discovered that demonstrate the power to produce large amounts of specified information. Premise two: intelligent causes have demonstrated the power to produce large amounts of specified information. Conclusion: intelligent design constitutes the best, most causally adequate, explanation for the information in the cell. It is the positive evidence in the second premise that makes this thinking logical.

Meyer presents an idea that I think all can live with. It is his notion that intelligent design is simply a conclusion that is the best explanation at present. This says nothing about how old the earth is, nothing about how animals and plants change now, and nothing about which religion is correct. If some evolutionists, like Richard Dawkins, can believe that aliens seeded life on this Earth,[6] then that is basically no different from saying that God did it. One is just as unscientific as the other. But as intelligent sources of life are vastly few and far between, and indeed interstellar space travel nigh on impossible, a god or God is certainly more plausible than cells put themselves together or aliens trucking it in.

I have some diagrams I use with my students that helps to illustrate this concept of ID and evolution. Look at Figure 18. What shape do you see? If you are like most, you'll say that you see a white triangle. News flash! There is no white triangle present. You are seeing what is not really there. Psychologists tell us that this is the property of closure in the brain. The brain looks for patterns and comes up with a shape seen in the past. It is sort of like the ancients creating the constellations in the night sky.

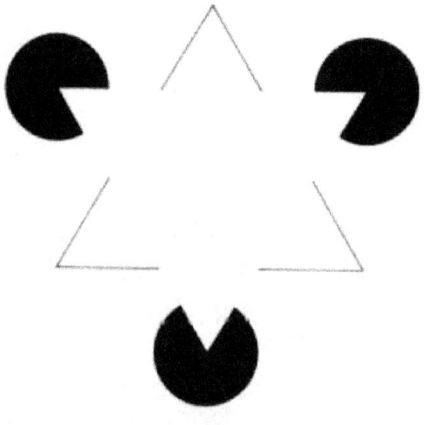

**Figure 18.**

Figure 19 is different. Note that while there are fewer shapes for each triangle, in both cases the white triangle is still present. This is how the evolutionists get you to see the triangle by eliminating any and all facts that would cause you to not see the triangle. Remember the Urey-Miller apparatus? Remember that no biology textbook will teach students about right- and left-handed amino acids? When evolutionists neglect to include vital information about something that in the long run will damage their case, they are involved with willful deception.

---

6. See clip of Richard Dawkins's interview in *Expelled*.

**Figure 19.**

Figure 20 shows a random pattern of shapes. I "created" this by cutting out the shapes from the first diagram, holding the pieces about three feet above the copying machine glass, and dropping them onto that glass surface. The truth is that two pieces fell white side down and one piece went onto the floor. I had to cheat to make randomness by turning the white sides up and re-dropping the errant piece. It fell onto the floor a second time. Hold the phone! Does this sound like an oxymoron to you? Can one cheat to make randomness? Does that tell you something about evolutionary experiments that "prove" Evolution Wrong true?

**Figure 20.**

## Is Evolution Compatible with Christianity?

Now direct your attention to Figure 21. This one gets the most giggles or smiles from my students. Responses usually include seeing a bird or a monkey. The funniest one heard to date is, "I see my sister." The point is that this is clearly designed. I had fun designing it. I played around with the angles and orientations of the eyes, horns, nose, and mouth. Ready for a teachable moment? Students are asked, if one drops the pieces one thousand times, will the face show up? one hundred thousand times? a billion times? It is here a hand or two might creep up, but not confidently. So how complex is this compared to a cell? They grasp the idea at once. They are told that they have a choice. Do you want to believe in something that is not really there (the white triangle=Evolution Wrong), or do you choose to believe that you were designed (the face diagram)?

**Figure 21.**

Here is the bottom line for this debate. Both sides can marshal evidence. Both sides have worldviews. Both have interpretations. Both sides make very little practical impact on operational (raise the standard of living, improve the lot of man) science. So the issue will rage on until God comes and then we will know who was right. Consider this though, if the evolutionists are right, it doesn't matter anyway as nothing matters if materialism only is true. If Christians are right, then God will judge the nonbelieving evolutionists. Christianity is a no lose proposition.

Return once more to the quote below the chapter's title. Sherlock Holmes almost always got it right because he simply followed the facts until the correct solution presented itself. True, he did not have to contend with his worldview when solving a case, but the key is to not ignore or explain away facts that cause problems with one's conclusion. Most evolutionists will do this to rescue the concept that Nature is all there is. Understand that if creation is true in any way, or if God is brought into the picture at all, then the worldview of that evolutionist will cease to exist. This will indeed make some of them very uncomfortable. As the Bible says in Hebrews 10:31, "It is a fearful thing to fall into the hands of the living God."

# Chapter 13

# Has Alien Life Evolved?

> "Life is infinitely stranger than anything which the mind of man could invent."
> —Sherlock Holmes, "A Case of Identity" (from *The Memoirs of Sherlock Holmes*)

Finding a Sherlock Holmes quote for this chapter was a bit of a challenge. After all, alien life on other planets was just not part of the lexicon or mindset during Sir Arthur Conan Doyle's time. That is not true today. Science fiction books, magazines, and films run the gamut of silly to serious. My own sci-fi beginnings were due mainly to two classic movies seen on television: *The Day the Earth Stood Still* and *Forbidden Planet* got me hooked.

But what does this topic have to do with the creation issue? Plenty. There are millions of stars in our galaxy alone. In the observable universe, there are one hundred to two hundred billion galaxies.[1] With the sheer number of stars in our galaxy alone, some people believe the odds are really good that life arose on some of the habitable planets of those stars, let alone the huge number of planets in the known universe.

Again the question goes back to, "So what?" The key here is that if an evolutionist can find life on another planet, then the odds are good that it formed similar to our planet's life had done. Or perhaps life is just starting,

---

1. Howell, "How Many Galaxies?"

and we can get good observational evidence of how evolution occurred on our world so very long ago.

That some are obsessed with finding life on other planets can be shown by where some are throwing their hard earned money. One such project is SETI (Search for Extra-Terrestrial Intelligence). It began two years after *Sputnik*, the first artificial satellite launched into space by the Russians. Two physicists at Cornell University published an article in *Nature* discussing how easy it would be to send messages via radio between the stars.[2] This led to a suggestion that searching for alien life could be done using radio telescopes.

The first actual listening for alien radio signals or transmissions happened on April 8, 1960. Frank Drake pointed a radio telescope at two nearby stars. For several weeks he scanned up and down the microwave band attempting to detect alien signals. Called Project Ozma, it was man's first attempt to locate an alien civilization.[3]

Soon other nations became involved. Almost one hundred projects were undertaken in the following years. NASA became interested in the work in the 1970s and finally became officially involved in 1988. It took until 1992 before serious work and observations could take place. One year after work began, Congress cut off all funding for the program. Today it continues, though at a much smaller scale, due to the generosity of private donors and foundations.[4]

Today, their hope is that by building their own SETI-dedicated radio telescope array, they will be free to use this device, and others like it in the future, to solely hunt for the elusive signals of alien intelligence. To quote from their own website, "It is possible that as telescope and SETI technology advance it may be possible to detect intelligence not by directed message but by the same kind of 'noise' we accidentally broadcast to the cosmos via radio, television and radar signals."[5]

*How will this project help evolution's cause?*

The group has indeed some ambitious projects and proposals. As an example, they are seeking to explain how planets form by themselves by simulating protoplanetary disk evolution. The reason for this is that one

2. "History of SETI."
3. "History of SETI."
4. "History of SETI."
5. "History of SETI."

## Is Evolution Compatible with Christianity?

cannot have life evolve on a planet if there is no planet to evolve on. They have developed their own science curriculum as well. One section, though, is geared solely to indoctrinating students with the standard evolutionary history of life on earth. The idea is that if the students understand how life evolved on Earth, they can then create life forms that can exist on another planet.[6]

I find this a very dubious undertaking. There are scientists with impressive credentials working on this very topic and getting nowhere. Differing scenarios have been proposed and all have come to naught. One simply reads Stephen Meyer's classic works such as *Signature in the Cell* and one can see why the ideas of man to create life are doomed to failure. So what makes the SETI group think that grade school children will solve the problem?

*I have heard that water is critical to life. Why is this so?*

If you are asking this question, you are only half right. It is liquid water that must be present in order for life to be possible at all. This is because molecules can dissolve in liquid water, allowing other complex biochemical reactions to happen. Another point in liquid water's favor is that it is crucial for the bending of enzymes.[7] For enzymes to work, they need an exact and certain 3-D shape. Water helps to have this happen.

Water, hereafter in a liquid form unless otherwise stated, is a most unusual substance. It has a very large temperature range for its liquid state. As most people know it is zero degrees Celsius to one hundred degrees Celsius. All other substances form liquids at temperatures far above or below surface temperatures of Earth. The only one that comes at all close is ammonia; however, it has a range of only thirty degrees Celsius.

One other interesting but nonetheless extremely important fact is that water is unique because it expands as it freezes. Because it does this, frozen water (ice) floats. This does not seem like much until one figures out that were water to sink upon freezing, soon every pond, lake, and ocean in cold climates would have no room for life as it would be filled with frozen water from the bottom up.

All these and many other facts point to the notion that water in its liquid form is the Holy Grail for planetary scientists.[8] Says Neil de Grasse

---

6. "LITU Curriculum Files."
7. Tyson, "Life's Little Essential."
8. Tyson, "Life's Little Essential."

## Has Alien Life Evolved?

Tyson, an astrophysicist and director of the Hayden Planetarium at the American Museum of Natural History, "Given that life on Earth is so dependent on water, and given that water is so prevalent in the universe, we don't feel that we're going out on a limb to say that life would require liquid water."[9] Now can you understand why evolutionists get so excited when water is discovered on Mars or elsewhere?

*But life requires a lot more than water. Where could we find "habitable planets"?*

To begin, planets that are outside of our solar system, habitable or not, are termed exoplanets. These objects are a rather new field of astronomy. However, astronomy, the actual science of the stars, has had a much longer history. Ever since man has gazed into the night sky, he has wondered about those "twinkle, twinkle little stars." Their constant positions in relation to one another gave us the constellation patterns we know today.

Soon people noted that there were a few stars that appeared to wander into and out of the constellations. Once telescopes were invented, these wanderers were found to be actual worlds of their own. Galileo's telescope showed the moons of Jupiter changed positions as they revolved around that world. So thus the planets must orbit the sun. But what of the planet we lived on? How did our world fit in with those things in the sky, not to mention the sun and moon as well?

Most books on evolution contain some reference to that contentious time when many, including the scientists of those days believed the earth was the center of the solar system as well as the entire universe, had their beliefs seriously challenged. The scientists of the time followed the teachings of Ptolemy, a Roman astronomer living around AD 100. The main contribution of his geocentric (Earth-centered) model was that it could explain why various celestial objects moved as they did. The first to upset this belief system was Copernicus in 1543. He still kept the solar system in the center of the universe, but now it was a heliocentric (sun-centered) model. This concept proposed was a radically different idea. According to one reference, it was so contentious that Copernicus did not have it published until after his death.[10]

After setting up this preface, the evolution texts will introduce Galileo as the hero of the story. Their premise is that the main foe of Galileo was

9. Tyson, "Life's Little Essential."
10. Iowa State University, "Copernican Model."

## Is Evolution Compatible with Christianity?

the church (Catholic Church in Rome). Galileo's persecution is set out as an example of today's science versus religion evolution/creation conflict. One example of this comes from Hansjörg and Wolfgang Hemminger's work hostile to creation ideas. They say,

> Today's Creationism . . . turns against the great Christian naturalists of the 15th and 16th century, against Copernicus, Galileo, Kepler and Newton. It repeats the proceeding against Galileo and argues in principle with the Inquisitors, for the issue at the trial was, among other things, whether the natural scientist had the freedom to set experimentation and observation above Scripture . . . Today's Creationists in principle have the same standpoint as the Inquisitors because they follow their empirical-biblicistic method.[11]

The problem is that they are wrong.

*But Galileo was persecuted, wasn't he?*

Of course he was. But his antagonist was not the church. Thomas Schirrmacher has written an excellent essay that fully documents these and other notions that lay to rest the Galileo versus church fairy tale. His foes were the scientists of his day who feared losing influence and scientific standing.[12] The truth also is that the Copernican system as well as Galileo's system were well regarded by church officials.[13] Schirrmacher writes,

> Until the trial against him, Galileo stood in high esteem among the Holy See, the Jesuits and especially the popes of his lifetime. His teachings were celebrated. Galileo's visit to Rome in 1611, after he had published his *Messenger from the Stars*, "was a triumph."[14]

Everyone is encouraged to investigate this momentous event in Galileo's life so that the truth will win out. Think about it—if the evolutionists are not concerned about the truth, then what about anything else they might say?

*Getting back to habitable planets, what exactly are those and how does this apply to the creation/evolution issue?*

---

11. Hemminger and Hemminger, *Jenseits der Weltbilder*, 201–2.
12. See Prause, *Niemand hat Kolumbus*, 182–83.
13. Schirrmacher, "Galileo Affair."
14. Koestler, *Sleepwalkers*, 426.

## Has Alien Life Evolved?

Some of the requirements have to do with the star the planet is orbiting. It must stay stable for billions of years. Of course, this is true only because evolutionists "know" that evolution takes this amount of time. It must not put out any intense radiations like those put out by pulsars. Those radiations would sterilize any planet orbiting it. The star must also not be a variable star, which is one that changes in brightness and energy output over time.[15]

Some of the requirements have to do with the planet. It must be in a system that has large outer planets to protect it from wandering asteroid bombardment, but not be too close to them either, which would upset the habitable planet's orbit. It cannot be too big and thus hold gasses such as hydrogen in its atmosphere nor too small and have the heavier gasses such as nitrogen and oxygen escape into space. It cannot be too close or too far away from the system's star. Most lists include lots of water be available. Interestingly, one list has that the crust must have plate tectonics working, which will keep mountains building while the weather erodes those mountains down.[16]

In looking at these lists, what is so obvious is that they are describing Earth. Of course they are, as we have life on it to the full. The big trick now is to find our twin out there. But this does not really answer the issue at all. Did life really come to be on this planet by the processes of Nature or by the hand of God as the Bible says? A natural science course at the University of Arizona asks the question, "Comparing with the other planets, it is clear that rather special conditions on Earth make life possible here! If all these conditions are met, is life inevitable or does it require something else? Or are we being too restrictive in our ideas, too tied to our particular forms of life and their requirements?"[17]

*Just how many exoplanets have been found and how many are habitable?*

The first one was found in 1994 by a radio astronomer Dr. Alexander Wolszczan. Wolszczan made his discovery by observing regular variations in the pulsar's rapidly pulsed radio signal, indicating the planets' complex gravitational effects on the dead star.[18] While some may have disputed his finding, all agree that any planet orbiting a pulsar would not be habitable at all.

15. "What are Requirements?"
16. "What are Requirements?"
17. Rieke, "Properties of Planets."
18. Brennan, "Expolanets 101."

## Is Evolution Compatible with Christianity?

One had to wait until 1995 to find the first exoplanets orbiting a star like our sun. Remember that for life, as we know it to exist, having a sun like ours would be a very important requirement. Another important factor is the development of new technologies. For example, NASA launched the Kepler Mission in 2009. Its sole purpose is to search for exoplanets. As of this date there have been over two-thousand of them confirmed.[19]

The problem is that very few of them are what we would call habitable. There is Kepler-69c, which is 1.7 times the size of Earth. The problem is that it is just inside the close to the star edge of the habitable zone. The conclusion is that it is more like Venus, and we know just how habitable Venus is.

Another key factor is to determine the atmospheric composition. This is as yet a technology not developed. Right now the James Webb Telescope is being developed for this very job. The drawback is that Earth-sized worlds would be too small to be detected. This means we will have to await the new machines capable of dividing the earth-like ones from those like Venus.

*We hear a lot about panspermia as a way of explaining how life came to Earth and it sort of has to do with life on other planets. Explain how panspermia works and why Christians ought to reject it.*

Panspermia is the idea that life originated from outside the earth and came to our planet via meteorites or comet dust particles. It has been around since the time of Lord Kelvin, who came up with the idea, but not the term. That honor belongs to Svante Arrhenius who invented the term in 1908.[20]

While it is true that Martian meteorites do exist, they are very rare indeed. Add to this the opinion of M. J. Burchell of the Centre for Astrophysics and Planetary Sciences at the University of Kent (United Kingdom), who says "given their size and transfer times (estimated from exposure to radiation in space), all will have received a sterilizing radiation dose during their transit to earth."[21]

Knowing this, it is even more doubtful that life came from Jupiter's moons or hitched a ride on comet dust. As Dr. Georgia Purdom of Answers in Genesis states in the sited article,

> The transfer of material from Mars, Jovian moons, and comets is plausible and in some cases has been documented. However, dust and rocks are not affected by the extreme cold and radiation of

---

19. Brennan, "Expolanets 101."
20. Purdom, "Did Life?"
21. Burchell, "Panspermia Today," 73–80.

## Has Alien Life Evolved?

outer space, whereas life would be and would probably not survive the journey to earth. Since life has not been found to exist in outer space it is doubtful life was transferred to earth from these locations.[22]

Related to panspermia is its cousin "directed panspermia." This idea is that life was seeded on Earth by an advanced alien civilization. Francis Crick, one of DNA's discoverers, and Leslie Orgel created this concept in 1973.[23] They further explain that since the earth is relatively young compared to the much older universe, other civilizations could have come and gone. Perhaps one of these felt it was their mission to seed habitable planets with bacteria so that evolution could take its course.

*Are there any other problems with this panspermia idea?*

There is one for panspermia you can read[24] about and another can be suggested for the notion of alien life or advanced civilizations. The notion of panspermia is threatened in a way by the very idea it was supposed to rescue—namely, evolution. One of the key ideas of evolution is natural selection, and it ought to apply to these microbial space travelers as well. One would suppose that if this sort of thing has been going on for billions of years, by now the travelers ought to be just about everywhere. Thus the problem—we don't find them *anywhere*.

Columbia University astrobiologist Caleb A. Scharf explains, "There are all sorts of plausible reasons. The simplest is that we've not yet managed to look very hard in all these places. It's also possible that we've just not put two and two together while studying the properties of terrestrial extremophilic organisms."[25] There is another possibility which, oddly enough, gives a nod toward creationists. Scharf concludes,

> But suppose we keep looking hard and find nothing—this would argue strongly against the possibility of galactic panspermia at all. And this would be interesting, because it would also serve to place a limit to the true extremes of life, a physical and chemical boundary condition. Perhaps the root cause turns out to be gravitational dynamics (interstellar transfer may be horrendously inefficient), or just the environmental limits of biochemistry and

22. Burchell, "Panspermia Today."
23. Crick and Orgel, "Directed Panspermia."
24. Scharf, "Panspermia Paradox."
25. Scharf, "Panspermia Paradox."

the molecular machines at the core of it all. In either case, a null result might actually tell us something vitally important about the phenomena of life, and our own cosmic significance.[26]

It is this key phrase at the very end where a creationist can hang his hat. The phrase "our own cosmic significance" echoes what the psalmist writes in Psalm 8:3–8.

> When I look at your heavens, the work of your fingers,
> > the moon and the stars, which you have set in place,
> what is man that you are mindful of him,
> > and the son of man that you care for him?
> Yet you have made him a little lower than the heavenly beings
> > and crowned him with glory and honor.
> You have given him dominion over the works of your hands;
> > you have put all things under his feet,
> all sheep and oxen,
> > and also the beasts of the field,
> the birds of the heavens, and the fish of the sea,
> > whatever passes along the paths of the seas.
> O Lord, our Lord,
> > how majestic is your name in all the earth!

That Caleb A. Scharf is a creationist is not what I believe or am trying to convey at all. I am simply saying that here is a scientist who is stating an interpretation that a biblical creationist can say; at least he is admitting we just might be the only intelligent life in the galactic neighborhood we call our own, not to mention the fact of life itself.

*What should a biblical Christian say about the idea that God could have created life, or even intelligent beings such as us, elsewhere?*

We must always be very careful when considering what God *could* do with what he *did* do. Obviously, God can do anything. The most bizarre idea out there is that God could have created you and everything in the universe two hours ago and also created you with false memories of events that never really happened. The big problem with this nonsensical idea is that what kind of god is that anyway? Certainly he is not described that way in his word. Nor does he present himself to be any type of trickster or charlatan. If he were, then nothing at all would be serious or true, as that would make the death and resurrection of Jesus to be false, the devil to be a pretend

---

26. Scharf, "Panspermia Paradox."

idea, and you can do the rest to prove to yourself the utter insanity of such a thought.

Now let us consider the notion that God does not tell us everything about everything in his word. That is indeed correct. He tells us exactly what we need to know and no more. He does not keep on babbling to hear himself talk. To know that he is true and that his word is true is to simply consider that every single event or prophecy ever written down in the Old and in some of the New Testaments has come to pass. That ought to make one confident of the remaining few prophecies found in the last sections of the New Testament.

This brings us to the last notion that oddly enough was echoed by a *Star Trek* TOS episode entitled "Bread and Circuses." In that episode on a planet that had "parallel evolution," Rome never fell and was the dominant civilization. The outsiders mentioned that they followed and worshiped the sun. At the end of the program, Kirk finds out that it is not the sun they worship but the Son of God.

Consider what the Bible says about Jesus, that he was slain *once* and for all (Heb 10:10: "And by that will we have been sanctified through the offering of the body of Jesus Christ once for all.") Also consider the verses that say *all* creation groans in the bondage of sin. Therefore, sin is universal. (Rom 8:22: "For we know that the whole creation has been groaning together in the pains of childbirth until now.") This means that if there is alien life and if it is a sentient being like ourselves, then it is sinful and in need of salvation. This also means that Jesus would not have had to die again on that planet *Star Trek* portrayed.

A final thought here that was the basis of a *Twilight Zone* episode. Advanced aliens came to Earth. One of their books in their language was studied by one of the show's main characters. About halfway through the show, the title is finally worked out. It was *How to Serve Man*. At the end, while many humans are boarding the ship, the other main character hears his friend calling to him just before he is forced into the ship. His friend calls out that the book *How to Serve Man* is a cookbook. Search for alien life? What if they are hungry?

# Chapter 14

# The Word of God and This Issue

> "This matter cuts very deep, and though I have not finally made up my mind whether it is a benevolent or a malevolent agency which is in touch with us, I am always conscious of power and design."
>
> —Sherlock Holmes, "The Hound of the Baskervilles"

This last section of investigation ends "In the beginning" with the word of God. Some of you have been taught the Scriptures from early on, some may have only recently come to faith in Jesus, and a few of you may be reading this book as an atheist or agnostic. Evolutionists, theistic or not, are not the best people to tell you what Christians or the Bible actually teaches. It is best to hear that information from credible sources.

There are two authors that I recommend. Both were skeptics and actually tried to prove the Bible wrong, or at least no better than any other religious text. This makes them credible, as they both know what it is like to be hostile to God's word. One is Josh McDowell, and the other is Lee Strobel. You would do well to start here as they both write clearly so that anyone can understand the questions and answers, regardless of your level of biblical understanding. It is true that there are others who have written critiques of these two. But a thought here—can everyone be right? No. Can everyone be wrong? Yes. Could only one of them be correct? Of course. But

## The Word of God and This Issue

these are not the critical questions to ask. The two key questions are: (1) Is the Bible historically reliable? (2) Is the Bible the word of God?

Some may object here by stating that it is disrespectful, if not even dangerous, to question the Bible. If we do so, are we not questioning God? First, let it be known that the questions asked here are not disrespectful at all. It is one thing to ask about reliability. It is a totally different thing to ask questions that are inane or do not shed light on a topic or issue. Besides, are we not to "taste and see that the Lord is good" (Ps 34:8)? When we examine the reliability of the Bible, it is like checking the freshness date on a carton of milk. In 1 Peter 2:2–3, the Bible is indeed compared to mother's milk, which is needed to nourish her infant child. Finally, if the Bible is what it says it is, the very words of God (1 Thess 2:13), there is no reason to fear. Does one really expect to find that God has written a book with errors in it?

When Paul and Silas went to the town of Berea (Acts 17:10) and preached about Christ in the synagogue there, what did the people do when they heard them? They not only received the message of God's grace found in Jesus, but they "examined the Scriptures every day to see if what Paul said was true" (Acts 17:11). You might respond that that is all well and good, but does not answer our concern about reliability. True, but it is a starting point. Paul (and God) commended the Bereans for seeking out the truth. Jesus himself calls God's word truth in John 17:17. Truth is never afraid of being examined. It is false ideas and notions that cower in dread when investigated. So let us join Benjamin Warfield, who said, "By all means let the doctrine of the Bible be tested by the facts and let the test be made all the more, not less stringent and penetrating because of the great issues that hang upon it."[1] Issues like the fact of creation, the need for a Savior, and the fact of an empty tomb hang in the balance.

It is easier to start the case for the New Testament being reliable as it is the more recent of the Bible's parts. Chauncey Sanders tells us that there are several tests used to verify the authenticity of an ancient text. The tests are: (1) bibliographical, (2) internal, and (3) external.[2]

These are rather thorough tests. Some, like bibliographical, have at least two parts to them. The first part of this test is to determine the time gap from the originals, or autographs, to the copies we have today. Skeptics and believers alike are sometimes surprised to find out that the two earliest copies of the New Testament date back as far as around AD 325. One is

---

1. Warfield, *Inspiration and Authority*, 217.
2. Sanders, *An Introduction*, 143.

## Is Evolution Compatible with Christianity?

the Codex Sinaticus and the other is the Codex Vaticanus. We can go back even further than that by reading the early church fathers, who quoted the New Testament books and verses so often in their writings that we could re-create the New Testament intact without even using any manuscripts.[3]

The bibliographical test's second part is to count how many manuscripts we actually have. Manuscript authorities routinely do this as part of their work, and bias does not come in to play here, for all they do is count. Now, if skeptics of the Bible accept the counters when they count the works of say Aristotle as being reliable, we only have five copies. The earliest copies, by the way, are fourteen hundred years after Aristotle lived. So the New Testament copies were over one thousand years closer to the originals. Add to this that the counters have counted over twenty-four thousand manuscript copies of portions of the New Testament. To sum it up, we can quote Sir Fredrick Kenyon, who was the principle librarian for the British Museum. "The interval, then, between the dates of the original composition and the earliest extant evidence becomes so small as to be in fact negligible, and the last foundation for any doubt that the Scriptures have come down to us substantially as they were written has now been removed."[4] If the evolutionist holds the secular figures of history to be true based upon such scant evidence, then why does he not accept the Bible based upon the sheer volume of evidence? Christians know the answer well. In John 3:19, we read, "This is the verdict: Light has come into the world, but men loved darkness because their deeds were evil."

The internal test has to do with the manuscripts themselves. This is a test to see if they are self-consistent or if they contradict themselves. One point made consistently is that the Gospel writers were actually witnesses to the events at the time of Jesus. In 2 Peter 1:16 we read, "For we have not followed cleverly devised tales when we made known to you the power and coming of our Lord Jesus Christ, but we were eyewitnesses of His majesty." Luke was not an original disciple of Jesus, but he was a doctor and as such knew the value of being there when things happened. He writes in Luke 1:1–2, "Inasmuch as many have undertaken to compile an account of the things accomplished among us, just as those who from the beginning were eyewitnesses and servants of the Word have handed them down to us." As you study the Bible, you will see that the writers support the evidence of other writers.

3. Geisler and Nix, *A General Introduction*, 357.
4. Kenyon, *Bible and Archeology*, 288–89.

## The Word of God and This Issue

The final test is the external test. This one looks to any other historical documents that will confirm or deny the internal testimony of the writings in question. One early church father was Papias, who was a disciple of John and was the bishop of Hierapolis close to the year AD 130. He wrote,

> This also the presbyter [the apostle John] said: Mark, having become the interpreter of Peter, wrote down accurately, though not in order, whatsoever he [Peter] remembered of the things said or done by Christ. For he neither heard the Lord nor followed him, but afterward, as I said, he followed Peter, who adapted his teaching to the needs of his hearers, but with no intention of giving a connected account of the Lord's discourses, so that Mark committed no error while he thus wrote some things as he remembered them. For he was careful of one thing, not to omit any of the things which he had heard, and not to state any of them falsely.[5]

Make time to research early church history. Note the line or succession of bishops. They are all in direct lines to either Peter or John. For example, Irenaeus was a disciple of Polycarp, who was a disciple of John. In one of Irenaeus's letters to a friend, it is said that Papias was also a disciple of John and a companion of Polycarp.[6]

So far we have put to rest any doubts one may have against the reliability of the New Testament. Therefore, we can now handle the issue of the Old Testament's reliability. We can start by noting that during the early church, there was no New Testament. All Paul and the disciples had was the Old Testament. So whenever Jesus quoted the Bible, it was the Old Testament. Seeing as the New Testament is reliable, therefore any Old Testament verses written in it are also reliable. That is at least a start.

Truthfully, we do not have the rich manuscript numbers for the Old as we do for the New. But what we do have are some factors that are just as good. First, one must take into account the incredible care that was taken when Jewish scribes would make a copy of the books of the Old Testament. There were extreme regulations: the scribe would have to wash his whole body, wear full Jewish clothes, have a certain number of lines after each chapter, and the ink had to be of an exact recipe. To find out more about this, research Samuel Davidson's book *The Hebrew Text of the Old Testament*. These scribes did their work from about AD 100 to about AD 500.

---

5. Eusebius, "Church History 3.39," 172–73.
6. Eusebius, "Church History 3.39," 170.

## Is Evolution Compatible with Christianity?

One of the main reasons we do not have many ancient manuscripts for the Old Testament is that rabbis gave greater honor to the newest copies. When a "Bible" got old, it had to be disposed of. There was no recycling in those days. So the proper way to get rid of it was to burn it or bury it.[7]

Until 1947, the oldest Old Testament we had was dated to AD 900. It was the discovery of the Dead Sea Scrolls that was the game changer. The scrolls amounted to about forty thousand inscribed fragments and were stored there by the librarians of the ancient Qumran community. More than five hundred books have been recovered, including every book of the Old Testament except for the book of Esther.[8]

The importance of this find cannot be overstated. It is the old age of these scrolls that is one of the two key reasons for their importance. They range from 200 BC to AD 68. Do the math. This gives us close to one thousand years from what we had before 1947. The second reason is that there is virtually no difference between the texts we had before 1947 and the Dead Sea Scrolls. Imagine having no copy machines to do your work. How would you stack up to the challenge of copying a book without any errors? F. F. Bruce sums up this evidence nicely when he says, "The new evidence confirms what we had already good reason to believe—that the Jewish scribes of the early Christian centuries copied and recopied the text of the Hebrew Bible with utmost fidelity."[9]

We come now to another avenue of evidence that Christians can use to convince skeptics that the writings of the Old and New Testaments are reliable. One can have science help. How can science help with religious questions? Perhaps this story will shed some light. A teacher was sitting next to me during a general session at the state science teacher's conference. The speaker allowed us to start a dialogue about his discussion points. During the sharing, he found out I was a creationist. He sarcastically asked which creation myth I would be teaching: the Native American, the Babylonian, or Norse creation myth? I told him that the one found in the Bible would be taught. He asked why, and was told that the Bible is backed up by the science of archeology. He got up and walked away. Perhaps he couldn't handle the fact of the Bible having scientific backing. Some might say that archeology is not considered a science. How much more, then, would paleontology not count as a science?

7. Bruce, *Books and Parchments*, 116.
8. Earle, *How We Got*, 50.
9. Bruce, *Second Thoughts*, 61.

# The Word of God and This Issue

For years, critics have discounted the Bible. If archaeologists hadn't discovered the physical remains of a city mentioned in the Bible, critics then claimed that the Bible couldn't be true. After all, they wanted physical proof. However, many ancient cities mentioned in the Bible have been excavated.

Consider the ancient city of Jericho. One of the recent excavators, the late Dame Kathleen Kenyon, stated that there was little to conquer by the time Joshua had gotten there, therefore she had stated that the story of the "Fall of Jericho" as told in Joshua 2–6 never really happened. Thanks be to God for archeologists and other scientists who do not accept the majority party line of what is true in their field, like Bryant Wood, archeologist from the University of Toronto. Gathering an eclectic array of evidence including pottery shards, Egyptian royal scarabs, and even a carbon-14 date, he has proven that the city walls really did come tumbling down and it was precisely at the time when Joshua would have entered that region of the Near East, and not 150 years after the city had been destroyed by an earthquake as claimed by Kenyon.[10]

Not to admit defeat, skeptics went after the second major claim made by Christians—that the Bible is the word of God. They claimed that the Old Testament, especially Genesis, was written from other ancient texts and changed to fit the monotheistic ideas of the Hebrew people. In response, one can say that it is also very possible the Hebrews had it first and other peoples perverted it into their way of understanding, as they did not want to follow the only true God. Look at early pagan beliefs. All involve some beings reflecting nature or super powered people. All have a prominent place for the sun. Now study Genesis. The sun is a created thing and not around until day four, so how could the Hebrews have borrowed from the pagans?

Another point in favor of the Bible not being man's word but that of God has to do with the fact that this is a collection of historical books that are not heroic epics. Note that the biblical histories do not hide the warts, failings, and sins of the people. There was and is only one unflawed person in the whole Bible, and that is Jesus. No other religion is so brutally honest of its heroes.

A third point that is evidence of God being behind the writing is the aspect of prophecy. Jesus' birth, ministry, even his death and resurrection, fulfilled three hundred prophecies. There were other historically accurate

---

10. Wood, "Walls of Jericho."

predictions, too, like how Alexander the Great's army conquered the city of Tyre. Read Ezekiel 26. Note how God reveals to the prophet the name of the city and how it will be destroyed. Verses 7–12a relate to King Nebuchadnezzar's siege of Tyre. Verses 12b–14 describes what happened when Alexander the Great came through that part of the world.

> "Also they will make a spoil of your riches and a prey of your merchandise, break down your walls and destroy your pleasant houses, and throw your stones and your timbers and your debris into the water. So I will silence the sound of your songs, and the sound of your harps will be heard no more. I will make you a bare rock; you will be a place for the spreading of nets. You will be built no more, for I the Lord have spoken," declares the Lord God.

He saw an island city, still known as Tyre. He was out to conquer the world, and no island city was going to say he could not get to them and be immune from his quest. Therefore, he ordered that the rubble that lay about from the coastal city of Tyre be thrown into the sea. He then marched across the rubble bridge and wiped out the island city. Remnants of this causeway are still visible today.[11]

A Bible scoffer I dealt with on a blog tried to say that Tyre has been rebuilt and therefore the Bible is wrong, as the prophecy was wrong. If one looks at the present site of Tyre, it is not on the original site. Plus if you go to the island city of Tyre, which was still called Tyre in those days, you will find a pile of stones and debris. The city (island) of Tyre has never been rebuilt so the prophecy of God still stands true.

There are other historically accurate prophecies found in Scripture. Isaiah foretold King Cyrus of Persia two hundred years before his birth. "It is I who says of Cyrus, 'He is My shepherd! And he will perform all My desire.' And he declares of Jerusalem, 'She will be built,' and of the temple, 'Your foundation will be laid'" (Isa 44:28).

Daniel predicts the rise and fall of four empires: from Babylon to Medo-Persia to Greece to end with Rome. He then told of how the Greek empire would be divided into four parts once Alexander the Great died and his four generals split up the conquered areas. How much more does the unbeliever want? The Bible is like no other holy book.

There is one more fascinating fact that points to the bible being God's word and not the words of men. Consider the notion of medically correct ideas of the nineteenth and twentieth centuries practiced by the Jews in BC

---

11. Butt, "Tyre in Prophecy."

## The Word of God and This Issue

times. How would they know to do this? In fact, they were the only group of people to practice such things in the years from the exodus onward.

William J. Cairney published a series of essays entitled "Biomedical Prescience" in the collection called *Evidence for Faith—Deciding the God Question* (edited by John Warwick Montgomery) that deals with this very topic. In the first volume, Cairney discusses what Moses tells the Israelites about clean and unclean animals. Reading this essay, we come to find out that pigs are "unclean" as they are breeding grounds for a host of worms and bacteria. Other unclean animals include rodents, bottom-feeding crustaceans, and animals with paws, such as dogs.

There are other directives Moses gave the people concerning farming. Why did they practice the "no crops in a field on the seventh year" method of crop rotation? This would rid the fields of various crop fungi and diseases. They also practiced what is called green manuring. This is where the plant stems and leftovers get plowed back under instead of using animal waste as fertilizer. Just how did the Jews become able to practice such healthy methods of farming and eating before the advent of modern knowledge of such safe practices?

Some might say that they got all this knowledge from the Egyptians while they were in captivity. A cursory study of Egyptian practices is found in S. I. McMillen's work, *None of These Diseases*. He lists some of the preparations used by the Egyptians as follows: crushed donkey teeth, rattlesnake fat, blood of worms, fly excrement, and manure from a variety of animals.[12] Obviously Moses' instructions did not come from any Egyptian living during his time. To find out where Moses got his instructions, all one has to do is read Exodus or Leviticus. For example, read Leviticus 11:1-2, "Now the Lord spoke to Moses and Aaron, saying to them, 'Speak to the children of Israel, saying, "These *are* the animals which you may eat among all the animals that *are* on the earth."'" Time and again when you read the writings of Moses, you will find that God does quite a bit of talking to Moses or Aaron. So just what more will it take for the unbeliever to see that the divine authorship of the Bible is not fantasy but a logical deduction from the evidence?

Getting back to the main topic of this book, we are now ready to apply God's word to this issue of creation/evolution. The following is based upon a Bible worksheet I use in my biology classes. The verses are chosen to show that evolution is totally incompatible with one who holds the Bible to be

---

12. McMillen, *None*, 11.

## Is Evolution Compatible with Christianity?

God's word. Before we get into the study, let us begin with how one of my previous pastors described or defined the Bible.

Before he began to read any verse(s) from the Bible used for his sermon, he said something like this, "We read from the authenticated, authoritative, inspired, inerrant, holy and true Scriptures which are the very words of God." Authenticated means that scholars have logically shown it to be legitimate. For example, liberal theologians claim the prophecies of Daniel were written after the events had taken place, thus explaining their accuracy. Proper studies have concluded that this is not the case, and that the antiquity of Daniel puts it well before the nations mentioned ever came to world prominence.

*Authoritative* has the root word of *authority*. When disputes arise or questions need answering, most will ask an expert or let some entity have the final say. Applied to the Bible, Christians say that when God speaks, we ought to listen (apologies to E. F. Hutton). Now today's secular person will say that we are following a terribly old-fashioned, out of date moral code and that we need to be "enlightened" so we can have more fun. So let's examine that idea of "fun." God says do not have sexual relations outside the bonds of marriage. If what the skeptics say is true, then their idea of fun is getting a disease like AIDS or hurt feelings when couples break up. If it is so much fun to be liberated, why is there so much misery?

Another tack to take when dealing with old-fashioned morality is to then ask the skeptic for his wallet. He will say that it is stealing if you take his money. Oh, that is such an old-fashioned concept you tell him. Start discussing the idea of murder being outdated and see the fur start to fly. Interesting that only the ideas about sex are old-fashioned but all the rest of the commandments are fine to stay. One thing I have always wondered is how does an eternal God write a book that goes out of date?

The doctrine of inspiration is not hard to understand. Moses, the prophets, and the New Testament writers were "carried along by the Holy Spirit" (2 Pet 1:16–21). Here is what one study Bible says about this section:

> This section is a strong statement on the inspiration of Scripture. Peter affirms that the Old Testament prophets wrote God's messages. He puts himself and the other apostles in the same category, because they also proclaim God's truth. The Bible is not a collection of fables or human ideas about God. It is God's very words given *through* people *to* people.[13]

---

13. *Life Application,* 2113, emphasis original.

## The Word of God and This Issue

It is important to note that the Bible is different from other religious texts as Peter says that the apostles and, by connection, the prophets were eyewitnesses of these things.

Another section of commentary says this about verses 20 and 21, "Men spoke from God as they were 'carried along by the Holy Spirit' means that Scripture did not come from the creative work of the prophet's own invention or interpretation. God inspired the writers, so that the message is authentic and reliable. God used the talents, education, and cultural background of each writer (they were not mindless robots); and God cooperated with the writers in such a way to ensure that the message he intended was faithfully communicated in the very words they wrote."[14] Look up 2 Timothy 3:16 and see what is said about inspiration there. Because this is the inspired word of God, this Bible is useful for teaching, rebuking, etc. In fact, St. Paul commends the Thessalonians for something that no other New Testament church did. In 1 Thessalonians 2:13, Paul recounts their initial reception of the Scriptures in this way: "When you received the word of God, which you heard from us, you accepted it not as the word of men, but as it actually is, the word of God, which is at work in you who believe."

Having established that the Bible claims to be the word of God, we can see what else it says about evolution, creation, and the like. The worksheet questions numbered here are the ones mentioned before. Do the readings and think about the answers you might give. Find the exact part of the verse that answers the question. Sometimes it is at the start, sometimes at the end, or maybe it is even the entire verse that gives the truth you seek. God bless your study. When you are done, the answer key will be provided in the Appendix. No cheating and reading ahead. It is thinking Christians who are commended time and time again by God and his writers.

1. According to Ephesians 3:8–9, who made everything? How does this back up Genesis 1?
2. Before continuing, we need to look at the Bible. What does 1 Thessalonians 2:13 tell you about the Bible?
3. Why can you trust God? Read Hebrews 6:18.
4. Why can't you trust people? See Numbers 23:19.
5. How did we get the Bible anyway? Read 2 Timothy 3:16.

---

14. *Life Application*, 2113.

## Is Evolution Compatible with Christianity?

6. A big question deals with the word *day* in Genesis 1. Which two words are consistently used near the word *day* that indicate time?

7. Now examine Genesis 19:1 and 19:27. The two words are used again. What do they mean?

8. Go back to Genesis 1 again. What word is used before "day" each time?

9. The word "day" is used scores of times in the Old Testament. Why do you think others never question meaning of it in those cases?

10. Look at 2 Corinthians 4:6. How does this support Genesis 1 ideas?

11. Many who compromise the truth of God's word with worldly knowledge claim the Bible tells who did the creating, but not how he did it. Find out how by reading 2 Peter 3:5. After that, tell us what the word deliberately means.

12. Read Psalm 33:6, 9. Verse 6 tells us how God did it and verse 9 tells us how long it took for it to be created each time He spoke it into existence. How does this contradict evolutionary beliefs?

13. Explain how Exodus 20:8–11 supports a seven-day creation week. Look up the origin of the seven day week.

14. What are we warned about in Colossians 2:8? Which one does evolution belong to? Why?

15. Read Proverbs 1:7. How is the biblical creationist different from an evolutionist?

16. Read Jeremiah 8:9 and tell how evolutionists differ from those who hold the Bible as truth.

17. Read Jesus' words in Mark 10:6. How does Jesus defend the truth of creation?

18. In Ephesians 4:14–15, consider how evolutionary ideas are "like a wind blowing here and there."

19. According to Jesus in John 5:46–47 what happens to those who do not believe Moses' writings?

20. As biblical creation is the foundation for Christianity, what does Psalm 11:3 tell us the world will try to do?

# The Word of God and This Issue

21. Ultimately the question of origins is not a matter of science. Read Hebrews 11:3 again to see what it is based upon. Note the verse also says how God did his creating.

# Chapter 15

# Four Websites Worth Your Time

> "Come, Watson, come. The game is afoot."
> —Sherlock Holmes, "The Adventure of Abbey Grange"

This chapter begins with one of the most famous of Holmes' quotes. Now the ball is in your hands, dear reader. Be like Holmes and investigate. But one might ask where to begin? Websites and blogs are quite prolific for the evolution camp. It might be helpful, once you are firmly and deeply holding on to the truth God has revealed to you via the Holy Spirit working through the word of God, to begin to examine one or two of the evolution sites or blogs to see where you can poke holes into their logic or thinking. Rare are the pro-evolution ones run by individuals having high-class efforts for civil discourse and debate. Be ready for foul language. Also be ready for them to set up straw men and twist your words around to make you sound illogical. They will appeal to authority, but remember that their authorities are fallible human beings, and we appeal to the authority above all authorities, the Holy God who spoke and the space-time continuum sprang into existence. Therefore, it is good to balance the evolution sites with those that support creation and intelligent design.

The following will list and describe four websites I routinely visit. They each will provide you with ample information for your quest to hunt down and destroy the secular worldviews of those you meet. You also will be able

## Four Websites Worth Your Time

to reason with those who try to compromise what God has said and merge it with what the evolutionists say is true. One is only for intelligent design, one is a summarizing of articles and then commentary for each, and two are apologetic in nature with a solidly Christian push for science and the Bible.

The first website is "Evolution News and Views." It is the one primarily tied to intelligent design, or ID. At the top, the site has a major essay about some topic relating to evolution. To the right side is a window with various titled articles of general interest and sometimes discussion or comments are allowed for them or the main top article.

Often though, it is the left sidebar listing of articles that vary in length, which are the most interesting. Those articles run the gamut of topics from evolution to education, from astronomy to zoology and everything in between. You want the latest take on what the evolution side says about retroviruses? You can find it here. How about the real truth behind the latest court ruling concerning ID? They cover it as well. Articles are fully documented and there are even video clips so you can hear both sides speak for themselves. Some quotes used in this book came from this valuable site.

Next up is "Creation/Evolution Headlines." This site covers almost every conceivable topic applicable to this issue, including ID once in a while. Of greater value to me is that there is commentary of a scientific nature analyzing articles written by evolutionists. So go ahead and look up the articles discussed and see what a fair job is done. At times this site will refer you to others, such as "Evolution News and Views," so I have two commentators covering the same topic. One of the sidebars is unique in that Bible-believing, creation positing scientists of the past are featured each month. One wonders how far science would have really progressed were it not for these persons who used the Bible as the basis for beginning to delve into the mysteries of God's world to see if they could figure out the lawgiver's laws of nature.

Third up is the "Institute for Creation Research," or ICR. This is the website for one of the longest running creationist organizations in the United States. This site features one main topic each day for analysis with a diversity of archived topics available. Everything from astronomy, biology, fossils, age dating claims, and language development to the latest rage in the media is covered from a scientific and biblical standpoint. Other features would be devotionals, resources of all types, and even a link to their graduate program.

## Is Evolution Compatible with Christianity?

Two areas of interest are the "Science Essentials" blog and the "Evidence for Creation" sidebar. The blog is hosted by Dr. Rhonda Forlow and contains many activities that anyone can do to explore the mysteries of God's creation using science type labs. Some activities are grade specific such as kindergarten through fifth grade, while others could be done from junior high on up. This is a boon for homeschoolers. When one clicks on the "Evidence for Creation" tab, it opens up to more specific areas such as "Evidence for Truth" or "Evidence from Science." Each area then lists several articles to read which will detail the evidence for the topic chosen.

Finally there is the website for "Answers in Genesis" (AIG), arguably the largest and most attacked creationist organization in the world. There are blogs to visit hosted by AIG staff. There is a complete list of seminars and conferences given. The article list is varied and extensive, most written by AIG staff and most of those have earned doctorates. I would be remiss if I did not mention that this outfit has opened a premier creation museum and boasts a full-scale replica of Noah's ark as a major new attraction at a separate site.

My favorite item to find on this website would be the skeptic's letters, which are answered idea-by-idea or paragraph-by-paragraph. This gives any creationist ideas on how to respond to critics of creation from many angles like scientific, philosophical, or scriptural.

If you subscribe to their website or order materials from them, you will receive via email a monthly offering of a free download which is one of the videos sold by them. If that is not your liking, then you also can use the website to preview selected clips of almost any video they offer. I find this valuable in that you can "try before you buy."

Keep reading, studying, praying, and sharing with others. Go to conferences, subscribe to magazines, watch the videos, and arm yourself. Evolution Wrong is one major stronghold Satan uses to tear down Christians and Christ's church. Satan will lose in the end and so will all of his followers. It is up to us to rescue those blinded by Satan. Being conversant in creation is one way to bring light to a darkened and deceived mind. May the Holy Spirit guide you as you deepen your knowledge and understanding of this issue. May God guard your heart and mind from the deceitful schemes of those blinded by Satan. Finally, be encouraged to use what you have learned to further the spreading of the gospel until he comes again to take us home. May we all hear the Father say, "Well done, good and faithful servant." Amen!

# Appendix

# Study Questions and Answers

*1. According to Ephesians 3:8–9, who made everything? How does this back up Genesis 1?*

The last part of verse 9 says God created all things. This says the same thing as Genesis 1. Note that the doctrine of God as Creator is not just an Old Testament idea. Also note that Paul is addressing his audience who would have been confronted with the two major ideas of his day: stoicism and Epicureanism. The latter is typified by Lucretius (b. 98 BC), who wrote about origins in his book *On the Nature of Things*. He said this about the earth, "It remains, therefore, that the earth deserves the name of mother which she possesses, since from the earth all things have been produced" and "of herself she created the human race."[1] The stoics held that matter was eternal and that if there was a god, it was not above nature. Likewise, it is important to note that the early church fathers also held to Genesis being the truth about creation. Theophilus wrote, "On the fourth day the luminaries came into existence. Since God has foreknowledge, he understood the nonsense of the foolish philosophers who were going to say that the things produced on earth came from the stars, so that they might set God aside. In order therefore that the truth might be demonstrated, plants and seeds came into existence before the stars. For what comes into existence later cannot cause what is prior to it."[2]

1. Lucretius, *On the Nature of Things* 5.795–796, 822–823.
2. Theophilus, *To Autolycus* 2.15.

## Appendix

*2. Before continuing, we need to look at the Bible. What does 1 Thessalonians 2:13 tell you about the Bible?*

The tail end of verse 13 says it is *not* the word of men, but the word of God. Another early church father was Basil. "Consider his words, Let us hear . . . the words of truth expressed not in the persuasive language of human wisdom but in the teachings of the Spirit, whose end is not praise from those hearing, but the salvation of those taught . . . The wise men of the Greeks wrote many works about nature, but not one account among them remained unaltered and firmly established, for the later account always overthrew the preceding one."[3] Does this sound like what truth is? If truth can get altered and over thrown, then it *is not* truth.

*3. Why can you trust God? Read Hebrews 6:18.*

Mid-verse says it is impossible for God to lie. Think about why we lie: (A) to avoid punishment; (B) to make ourselves look good . . . remember dating?; (C) to get what is not ours. Knowing these reasons to lie, does God even need to lie? Of course not! Think about this as well, if God lies, then why are you reading his book and saying things like the creeds in church? The creeds are all statements of faith. They are only as good as what they are based upon. As they are based upon Scripture, then if a liar wrote the Bible, then the creeds are not trustworthy also.

*4. Why can't you trust people? See Numbers 23:19.*

It states, "God is not a man (human) that he should lie." So people lie, and not just politicians, lawyers, and used car salesmen do this . . . all lie! (This includes evolutionists.) The early church fathers were somewhat kinder in describing the "wisdom" of the Greeks. Theophilus said that, "The world is not uncreated nor is there spontaneous production of everything, as Pythagoras and the others have babbled."[4] He also took on Plato and the Age of the Earth when he wrote, "If some period has escaped our notice, say 50 or 100 or even 200 years, at any rate it is not myriads, or thousands of years as it was for Plato . . . and the rest of those who wrote falsehoods."[5]

*5. How did we get the Bible anyway? Read 2 Timothy 3:16.*

---

3. Basil, *On the Hexameron* 1.1.2; Basil, *The Fathers of the Church* 46:4, 5.
4. Theophilus, *To Autolycus* 3.25, 28.
5. Theophilus, *To Autolycus* 3.29.

## Study Questions and Answers

It is "God breathed," which means inspired. The writers still had their own styles, idioms, etc., but these are *God's* words. This also means it is inerrant and a standard of absolute truth. The Institute for Creation Research website puts it this way,

"Given by inspiration of God" is all one word in the Greek, *theopneustos*, meaning "God-breathed." This word refutes any idea of human inspiration (such as a poet, or musician might claim). The Scriptures, by whatever particular methods God may have used in their various parts, including the individual human abilities and researches of the various human writers (whose abilities He had created and whose researches He had guided), as they finally came from their Spirit-guided minds and pens, are in effect God-breathed.[6]

*6. A big question deals with the word day in Genesis 1. Which two words are consistently used near the word day that indicate time?*

Evening and morning are used. Liberals and theistic evolutionists insist that these words mean beginning and ending of a vast time span. Really? Look at number 7.

*7. Now examine Genesis 19:1 and 19:27. The two words are used again. What do they mean?*

They mean the beginning and ending of one solar day.

*8. Go back to Genesis 1 again. What word is used before "day" each time?*

It is a number word (first, second, etc.). A Hebrew writer of that time had no other concept of vast ages when a number was used with the word *day*. He wrote and meant one twenty-four-hour day.

*9. The word day is used scores of times in the Old Testament. Why do you think others never question meaning of it in those cases?*

One never hears the evolutionists or theistic creationists discuss the word *day* when it occurs anywhere else in Genesis or the rest of the Bible, for that matter. The only time the word *day* cannot mean a twenty-four-hour day in their minds is in Genesis chapter 1. Those people reject sound Hebrew scholarship stating that there is no other meaning for the word *day* in Genesis 1 other than a twenty-four-hour day. The only reason they have

---

6. Institute for Creation Research, "New Defender's Study Bible Notes," https://www.icr.org/bible/2Timothy/3/16–17.

# Appendix

for rejecting it is that a regular solar day is not compatible with evolution in any way, shape, or form. This is why I say it is the beginning that is key to understanding this issue.

10. *Look at 2 Corinthians 4:6. How does this support Genesis 1 ideas?*

It basically is the same thing as "Let there be light." Here then is another New Testament verse that echoes Genesis 1. Note that Paul is making an analogy here. He is using the creation of light out of darkness to mirror what God has done for each of us who believe. God has caused his light to shine in our hearts that were once ruled by the darkness of sin. It is crucial that we never forget the constant battle we wage against that darkness. Just as day and night cycle every twenty-four hours, we "cycle" between sin and forgiving grace each and every twenty-four hours. Always remember that if Genesis 1 is a myth, then so is what Paul is saying here just another tale told around the fire to amuse us.

I wish to tell a true story related to this issue of God making light in Genesis 1. I had an experience with two men at one of my presentations at a non-public school conference. After the session had ended and most people had filed out, two gentlemen stood in the aisle, both with arms across their chest and grumpy looks on their faces. I asked if I could help them. One of them sneered, "That is just so ridiculous that God made light without the sun. How is that even possible?" I still believe that God inspires, and without skipping a beat, I walked over to the light switch, flicked it off, then back on, and asked, "We make light without the sun all the time. Why can't God?" The looks of scowls slowly turned into amazement. "I never thought of that" was the reply. I was tempted to respond back with a strong encouragement for them to start thinking.

11. *Many who compromise the truth of God's word with worldly knowledge claim the Bible tells who did the creating, but not how he did it. Find out how by reading 2 Peter 3:5. After that, tell us what the word deliberately means and explain the verse.*

The middle of the verse says that by God's word (speaking) the cosmos came into being. Do we know just how that works? No. Does science need to explain that? No. Remember that science can only explain the physical, and not the supernatural. Later in the verse the word *deliberately* appears. Deliberately means on purpose. This tells us the people of old forgot *on purpose*. Please realize that this means all have had a chance to know about

## Study Questions and Answers

God. In Romans, Paul tells us that everyone has evidence of God's existence from creation. Next comes the light of revelation. There are stories of unbelievers having dreams of Jesus coming to them saying that he has heard their prayers asking him to make God real to them. The truth is that unbelievers on judgment day are without excuse.

12. *Read Psalm 33:6, 9. Verse 6 tells us how God did it and verse 9 tells us how long it took for it to be created each time he spoke it into existence. How does this contradict evolutionary beliefs?*

In verse 6 we find that God did it by his word, verse 9 says he spoke, and it came to be. Consider the concept of "no time." God called for fish, and instantly fish in the actions of swimming appeared in all oceans and lakes. Same thing is true of birds. Were there a Blu-ray or DVD of that creation day, it would be a brilliant blue sky, verdant forests the likes of which we have never seen, and as the camera pans down to the shoreline we wou ... birds! Think about when Jesus healed people. There was no wait for natural processes. They were healed instantly. Find Mark 4:35–41. How much do you want to bet that not only was the sea dead calm after he commanded it to "be still," but also their clothes were dry? Read in verse 41 that they were terrified after this happened. Why would they be terrified if it took five hours for the storm to die down by natural causes? When the Lord of creation speaks, the storm listens and responds instantaneously.

13. *Explain how Exodus 20:8–11 supports a seven-day creation week. Look up the origin of the seven-day week.*

How great that we do not have a God who says do as I say but not as I do. Instead, God tells us to do as he did. "I [God] worked six days; you work six days." This commandment has no meaning if the days of Genesis 1 were not six twenty-four-hour days. The reference source you use to look up the origin of the seven-day week might say something about farmers bringing their goods to market every seven days, but this begs the question. Obviously, the source's author does not want to give any credence to Genesis 1, thus this "farmers" answer is an attempt to stifle the legitimate pondering and asking questions whose answers point to the Bible being true.

14. *What are we warned about in Colossians 2:8? Which one does evolution belong to? Why?*

## Appendix

Watch out for hollow and deceptive human philosophy. Evolution is said to be a science, but it has a philosophical side as well, which means that part is a totally human idea. In Genesis, we read that everything is commanded to reproduce after its kind. What do we see with fruit flies, bacteria, and thousands of years of animal breeding? Animals are reproducing after their kind. Of course there are variations. No Christian scientist denies this fact. What any biblically sound Christian must reject is the unbiblical ideas that life can put itself together, and that animals can change into totally different animals via mutations of DNA. Thus, macroevolution is not scriptural, and therefore it is man-made.

15. *Read Proverbs 1:7. How is the biblical creationist different from an evolutionist?*

We start with God's truth and a biblical worldview, so we have the correct foundation for all knowledge of whatever type (biology, psychology, etc.). Remember that if you deal with an atheist evolutionist, remind him that according to his worldview he cannot trust his thoughts or ideas as he has a mutant fish brain. Truth must be revealed from outside our system as we are in the system. It is also noteworthy that tigers kill tigers, so why is it wrong for humans to kill humans? His response ought to be interesting. Finally point out to him that *if* evolution is true, then nothing we do matters as it is either universal heat death or a Big Crunch that our existence will end. Either way *everything* goes back to nothing.

16. *Read Jeremiah 8:9 and tell how evolutionists differ from those who hold the Bible as truth.*

They have rejected the word of God and therefore have no wisdom. Ponder this: what was illegal in the past is now legal today. There are those who wish to legalize activities or substances that are presently illegal. With something like slavery, it is good we change bad law. But with areas like mind-altering substances, sexuality, and (when we can kill someone) abortion and euthanasia, would not moral absolutes be a wiser approach? Since God knows everything, why not trust him with what will be for your good and trust him to point out what would be wise (wisdom) to avoid.

17. *Read Jesus' words in Mark 10:6. How does Jesus defend the truth of creation?*

Male and female did not evolve, as they were that way "since the beginning." Note also that of all God's living things only humans were fashioned or made "by hand" as it were. All other living things were spoken into existence. "Let there be" is how God did it for them. This means that humans were the crowning glory of his creation. This makes us not animal at all. This is why he gave us dominion over his works. This is also why the Father sent the Son to die and rise again and the Spirit proceeds from them both to give us faith to believe in Jesus. That is because he loves us that much. Recall that in Genesis 3 when Adam and Eve fell into sin, God went looking for them as he felt the relationship had broken. Finally, consider that Jesus knew what he was talking about as he actually was there.

*18. In Ephesians 4:14–15, consider how evolutionary ideas are "like a wind blowing here and there."*

Those evolutionary ideas are always changing, just like the winds change, in order to accommodate new fossil discoveries that are problematic. They do not even know how evolution works. Consider that evolution is slow except when it is rapid. Animals like scorpions and coelacanths remain unchanged for literally hundreds of millions of years, but other animals underwent rapid changes and that is why we have no fossils for them. How about that we have no fossils for intermediate forms as macroevolution is bogus? Now are we saying that science must never change what it holds as true? If new data points to a new conclusion, then by all means change what used to be held true. But this is not what is done with evolution. It is presently so plastic as to mean anything and everything, which really explains nothing. If new data show it is wrong, then the idea is modified. Remember that the evolutionist cannot allow for the possibility that creation could be true. To do so would be to deny his god, materialism, is god.

*19. According to Jesus in John 5:46–47 what happens to those who do not believe Moses' writings?*

Accepting Genesis is foundational to God's salvation plan, but not a part of God's salvation plan. One needs to accept Jesus as Savior to be saved, but if Genesis is a myth, then so is sin. If sin is a myth, then why did Jesus have to die? Think about what Jesus says in Luke 16:31. He said to him, "'If they do not hear Moses and the Prophets, neither will they be convinced if someone should rise from the dead.'" So, if one doesn't believe Moses writings, he will not believe what he says either.

## Appendix

20. *As biblical creation is the foundation for Christianity, what does Psalm 11:3 tell us the world will try to do?*

It will try to destroy our foundation. They know if Genesis is gone, the rest of Scripture falls apart. This leads us to ask why is it that every other religion gets a pass in public schools except Christianity? Why is it that the ACLU rarely defends Christians' rights but will easily defend the atheists' rights? I submit that the answer for both questions is due to the fact that Christianity is right.

21. *Ultimately the question of origins is not a matter of science. Read Hebrews 11:3 again to see what it is based upon. Note the verse also says how God did his creating.*

It is a matter of faith. It is important to note that as evolution has no experimental evidence that: 1. life can create itself, 2. information can make itself (think DNA), or 3. that animals can change into completely different life forms. What we do have are plenty of observations that life comes only from pre-existing life, that intelligence is needed to create information, and that living things do change, but only in a very narrow range. Based upon these two lists evolution is a matter of faith also. The main idea is that biblical creation is a reasonable faith and the evolutionist's faith is unreasonable. Verse 3 states that God used his word to do his creating, just as other verses have stated.

Related to this last idea of "Word" is an old PBS program called "Firing Line" which ran televised debates between opposing sides of a question. There was one dealing with Creation, ID, and Evolution. Towards the end, Rev. Barry Lynn, an active leader in Americans for the Separation of Church and State, said that in John's Gospel is the first verse that reads, "In the beginning was the Word." He continued to say that word just might have been a command, "Evolve." Were the Rev. Lynn a minister with New Testament Greek training he would not have made that statement, at least if he were an ethical person. The Greek word that is used in that verse for "Word" is *Logos*. Logos has no meaning of evolve. So, Lynn was playing fast and loose with God's word. Just what kind of a minister is he? Certainly he will have to answer to Jesus for deliberately misleading people into thinking that Jesus, who is the living Word as the context of John chapter 1, demands we acknowledge him to be a word meaning evolve.

# Bibliography

American Geological Institute. *Investigating the Earth*. Boston: Houghton Mifflin, 1984.
Aquinas, Thomas. "On the Things That Belong to the Seventh Day." www.sacred-texts.com/chr/aquinas/summa/sumo82.htm.
Batten, Don, ed. *The Revised and Expanded Answers Book*. Green Forest, AR: Master Books, 2001.
Bergman, Jerry. "Scientists Urge Censorship of Terms Implying Design and Purpose When Describing Life." *Darwinism* (blog), *Answers in Genesis*, May 4, 2011, https://answersingenesis.org/charles-darwin/darwinism/scientists-urge-censorship-of-terms-implying-design-in-life/.
Brennan, Pat. "Expolanets 101." *The Search for Life* (blog), *Expolanet Exploration Program*, accessed March 28, 2019, https://exoplanets.nasa.gov/the-search-for-life/exoplanets-101/.
Bruce, F.F. *The Books and Parchments*. Rev. ed. Westwood, NJ: Fleming H. Revell, 1963.
———. *Second Thoughts of the Dead Sea Scrolls*. 2nd ed. Grand Rapids: Eerdmans, 1980.
Burchell, M.J. "Panspermia Today." *International Journal of Astrobiology* 3 (2004) 73–80.
Butt, Kyle. "Tyre in Prophecy." *Inspiration of the Bible* (blog), *Apologetics Press*, 2006, www.apologeticspress.org/apcontent.aspx?category=13&article=1790.
Clery, Daniel. "Impact Theory Gets Whacked." *Science* 342 (October 2013) 183–85. www.sciencemag.org/content/342/6155/183.
Collins, Nick. "Tyrannosaurus Rex 'Hunted in Packs.'" *Dinosaurs* (blog) *The Telegraph*, June 22, 2011, www.telegraph.co.uk/news/science/dinosaurs/8589113/Tyrannosaurus-Rex-hunted-in-packs.html.
The Commission on Theology and Church Relations. *In Christ All Things Hold Together: The Intersection of Science & Christian Theology*. St. Louis: The Lutheran Church—Missouri Synod, 2015.
Coppedge, James F. *Evolution: Possible or Impossible*. Grand Rapids: Zondervan, 1973.
Crick F.H.C. and L. E. Orgel. "Directed Panspermia." *Icarus* 19 (1973) 341–46.
Dejoie, Joyce and Elizabeth Truelove. "StarChild Question of the Month for October 2001." https://starchild.gsfc.nasa.gov/docs/StarChild/questions/question38.html.
DeYoung, Don. *Thousands . . . Not Billions: Challenging the Icon of Evolution, Questioning the Age of the Earth*. Green Forest, AR: Master Books, 2005.
Earle, Ralph. *How We Got Our Bible*. Grand Rapids: Baker, 1975.
Erwin, Douglas. "News Brief." *Geotimes* (February 1991) 32.

# Bibliography

Eusebius. "Church History 3.39." In *A Select Library of the Christian Church: Nicene and Post-Nicene Fathers: First Series*, edited by Philip Schaff, 172–73. Reprint, Peabody, MA: Hendrickson, 1994.

*Evolution: The Grand Experiment*, DVD. Directed by Carl Werner. Green Forest, AR: New Leaf Publishing Group, 2009.

*Expelled: No Intelligence Allowed*. DVD. Directed by Nathan Fankowski. Salt Lake City, UT: Premise Media Corporation, 2013.

"The Firing Line 1997 Creation-Evolution Debate." Debate, Seton Hall University, South Orange, NJ, December 4, 1997. www.biblicalcatholic.com/apologetics/p45.htm.

"First Creature on Earth Significantly More Complex." Presuppositions (blog), *Answers in Genesis*, April 12, 2008, https://answersingenesis.org/presuppositions/first-creature-on-earth-significantly-more-complex.

Fisher, Robert B. *God Did It, But How?* Ipswich, MA: American Scientific Affiliation, 1997.

Flores, Graciela. "Journals and Intelligent Design." *The Scientist* 19 (2005) 12.

Forrest, Barbara, and Paul R. Gross. *Creationism's Trojan Horse: The Wedge of Intelligent Design*. New York: Oxford University Press, 2004.

Fortner, Rosanne W. "Down to Earth Biology." *The American Biology Teacher* 54 (1992) 76–79.

Galilei, Galileo. *Dialogue Concerning the Two Chief World Systems, Ptolemaic and Copernican*. Translated by Stillman Drake. Berkley: University of California Press, 1962.

Geisler, Norman L. and William E. Nix. *A General Introduction to the Bible*. Chicago: Moody Press, 1979.

Ghose, Tia. "New T-Rex Tracks Add to Pack-Hunting Theory." Animals (blog), *Discovery News*, July 24, 2014, https://www.seeker.com/new-t-rex-tracks-add-to-pack-hunting-theory-1768850063.html.

Giberson, Karl. *Saving Darwin: How to Be a Christian and Believe in Evolution*. New York: HarperOne, 2008.

Gould, Stephen J. *Ever Since Darwin: Reflections in Natural History*. New York: Norton, 1977.

———. "The Evolution of Life on Earth." *Scientific American* 271 (1994) 84–91.

———. "Nonoverlapping Magisteria." *Natural History* 106 (1997) 16–22.

———. *Wonderful Life: The Burgess Shale and the Nature of History*. New York: Norton, 1989.

Ham, Ken, ed. *The New Answers Book: Over 25 Questions on Creation/Evolution and the Bible*. Green Forest, AR: Master Books, 2006.

———. *The New Answers Book 4*. Green Forest, AR: Master Books, 2013.

Hein, Steven A. "Reason and the Two Kingdoms: An Essay in Luther's Thought." *The Springfielder* 36 (1972) 138–48.

Hemminger, Hansjörg and Wolfgang Hemminger. *Jenseits der Weltbilder: Naturwissenschaft, Evolution, Schöpfung*. Stuttgart, Germany: Quell Verlag, 1991.

Hickman, Cleveland, et al. *Integrated Principles of Zoology*. 10th ed. Boston: McGraw-Hill, 1997.

"History of the SETI Institute." About Us (blog), *SETI Institute*, 2019, https://www.seti.org/history-seti-institute.

Hitchens, Christopher. "The Hitchens Transcript." Interview by Marilyn Sewell, *Portland Monthly*, December 17, 2009, https://www.pdxmonthly.com/articles/2009/12/17/christopher-hitchens.

# Bibliography

Howell, Elizabeth. "How Many Galaxies Are There?" *Science & Astronomy* (blog), *Space.com*, March 20, 2018, https://www.space.com/25303-how-many-galaxies-are-in-the-universe.html.

Hughes, Jennifer F. et al. "Chimpanzee and Human Y Chromosomes Are Remarkably Divergent in Structure and Gene Content." *Nature* 463 (January 2010) 536–39. https://www.nature.com/articles/nature08700.

Iowa State University. "The Copernican Model." Unit 2 (blog), *Polaris Project: Evening Star*, 2000–2001, www.polaris.iastate.edu/EveningStar/Unit2/unit2_sub2.htm.

Jablonka, Eva, and Marion L. Lamb. "Soft Inheritance: Challenging the Modern Synthesis." *Genetics and Molecular Biology* 31 (2008) 389.

Kenyon, Fredrick G. *The Bible and Archeology*. New York: Harper, 1940.

Koestler, Arthur. *The Sleepwalkers: A History of Man's Changing Vision of the Universe*. London: Hutchinson, 1959.

Lewontin, Richard. "Billions and Billions of Demons." *The New York Review of Books* 44 (1997) 31.

Libby, Willard Frank. *Radiocarbon Dating*. Chicago: University of Chicago Press, 1952.

*Life Application Study Bible*. Wheaton, IL: Tyndale House, 2005.

"LITU Curriculum Files." *Education and Outreach* (blog), SETI Institute, 2019, https://www.seti.org/education-outreach/life-universe-litu-curriculum-files.

Luskin, Casey. "Problem 6: Molecular Biology Has Failed to Yield a Grand 'Tree of Life.'" *Evolution News and Views*. Article published February 2, 2015. https://evolutionnews.org/2015/02/problem_6_molec/.

———. "Real Science vs. Bill Nye the 'Science' Guy." *Evolution* (blog), *Evolution News and Science Today*, February 2, 2015, https://www.evolutionnews.org/2015/03/bill_nye_resppono94591.html.

McMillen, S.I. *None of These Diseases*. Old Tappan, NJ: Revell, 1984.

Meyer, Stephen C. *Signature in the Cell: DNA and the Evidence for Intelligent Design*. New York: Harper One, 2009.

Miller, Kenneth R. *Finding Darwin's God: A Scientist's Search for Common Ground between God and Evolution*. New York: Cliff Street Books, 1999.

———. *Only a Theory: Evolution and the Battle for America's Soul*. New York: Viking, 2009.

Miller, Kenneth R., and Joseph A. Levine. *Biology*. Upper Saddle River, NJ: Pearson, 2010.

Mitchell, Elizabeth. "Truth from Telegraph, the World's Newest Zonkey." *Answers in Depth* (blog), *Answers in Genesis*, August 28, 2014, https://answersingenesis.org/hybrid-animals/truth-telegraph-worlds-newest-zonkey/.

Mitchell, Melanie. *Complexity: A Guided Tour*. Oxford: Oxford University Press, 2009.

Monastersky, Richard. "The Whales' Tale: Searching for the Landlubbing Ancestors of Marine Mammals." *Science News* 156 (1999).

Morris, Henry M. *Defender's Study Bible*. Nashville: World Publishing, 2006. www.icr.org/bible/Joshua/10:12–14.

Morris, John D. "Does 'The Beak of the Finch' Prove Darwin Was Right?" *Acts & Facts* 23 (1994).

———. *The Young Earth*. Green Forest, AR: Master Books, 2002.

Morris, Simon Conway. *The Crucible of Creation: The Burgess Shale and the Rise of Animals*. Oxford: Oxford University Press, 1998.

# Bibliography

Mortenson, Terry. "National Geographic Is Wrong and So Was Darwin." *Charles Darwin* (blog), *Answers in Genesis*, November 6, 2004, https://answersingenesis.org/charles-darwin/national-geographic-is-wrong-and-so-was-darwin.

National Geographic Society. "Mass Extinction." *Science & Innovation* (blog), *National Geographic*, accessed March 28, 2019, https://www.nationalgeographic.com/science/prehistoric-world/mass-extinction/.

National Science Foundation. "And the First Animal on Earth Was A . . ." *News* (blog), *National Science Foundation*, April 10, 2008, https://www.nsf.gov/news/news_summ.jsp?cntn_id=111408.

Nye, Bill. "Humanist of the Year Award." Speech. American Humanist Association, San Jose, CA, January 5, 2010, https://www.youtube.com/watch?v=S4dZWbFs8To.

Patterson, Roger. "Chapter 1: What Is Science?" In *Evolution Exposed: Biology*. Petersburg, KY: Answers in Genesis, 2007. https://answersingenesis.org/what-is-science/what-is-science.

Prause, Gerhard. *Niemand hat Kolumbus ausgelacht: Fälschungen und Legenden der Geschichte richtiggestellt*. Düsseldorf, Germany: Econ-Verlag, 1966.

Purdom, Georgia. "Did Life Come from Outer Space?" In *The New Answers Book 3*, edited by Ken Ham. Green Forest, AR: Master Books, 2010. https://answersingenesis.org/origin-of-life/panspermia/did-life-come-from-outer-space.

———. "Peppered Moths: The Saga Continues." *Blogs* (blog), *Answers in Genesis*, March 22, 2012, https://answersingenesis.org/blogs/georgia-purdom/2012/03/22/peppered-moths-the-saga-continues.

Rieke, G. H. "Properties of the Planets and Habitable Zones." http://ircamera.as.arizona.edu/NatSci102/NatSci102/lectures/habzone.htm.

Rokas, Antonis, and Sean B. Carroll. "Bushes in the Tree of Life." *PLOS Biology* 4 (2006) e352.

Rossiter, Wayne. *Shadow of Oz: Theistic Evolution and the Absent God*. Eugene, OR: Pickwick, 2015.

Samec, Ronald. "Lunar Formation—Collision Theory Fails." *Journal of Creation* 27 (2013) 11–12.

Sanders, Chauncey. *An Introduction to Research in English Literary History*. New York: Macmillan, 1952.

Sarfati, Jonathan and Michael Matthews. "Argument: The Fossil Record Supports Evolution." In *Refuting Evolution* 2. Powder Springs, GA: Creation Book Publishers, 2011. https://creation.com/refuting-evolution-2-chapter-8-argument-the-fossil-record-supports-evolution.

Scannella, John B., et al. "Evolutionary Trends in *Triceratops* from the Hell Creek Formation, Montana." *Proceedings of the National Academy of Sciences* 111 (July 2014) 10245–250.

Scharf, Caleb A. "The Panspermia Paradox." *Life, Unbounded* (blog), *Scientific American*, October 15, 2012, https://blogs.scientificamerican.com/life-unbounded/the-panspermia-paradox.

Schirrmacher, Thomas. "The Galileo Affair: History or Heroic Hagiography." *Creation* (blog), *Answers in Genesis*, April 1, 2000, https://answersingenesis.org/creation-scientists/the-galileo-affair-history-or-heroic-hagiography.

Schlossberg, Herbert. *Idols for Destruction: Christian Faith and Its Confrontation with American Society*. Wheaton, IL: Crossway, 1993.

# Bibliography

Sewell, Curt. "Carbon-14 and the Age of the Earth." *Essay-Links* (blog), *Revolution Against Evolution*, November 8, 1999, https://www.rae.org/essay-links/bits23/.

Singer, Peter. *Animal Liberation*. New York: Ecco Books, 2002.

Sloan, Christopher P. "Feathered Dinosaurs." *National Geographic* 197 (November 1999) 98–107.

Snelling, Andrew. "Where Are All the Human Fossils?" *Creation* 14 (December 1991) 28–33.

Stallard, Brian. "Whales Kept Their Hip Bones for Better Sex." *Animals* (blog), *Nature World News*, September 9, 2014, https://www.natureworldnews.com/articles/8948/20140909/whales-kept-hip-bones-better-sex.htm.

Stambaugh, James. "Death Before Sin?" *Arts & Facts* (blog), *Institute for Creation Research*, May 1, 1989, https://www.icr.org/article/295.

Stanovich, Keith. *The Robot's Rebellion: Finding Meaning in the Age of Darwin*. Chicago: University of Chicago Press, 2004.

Technische Universitaet Muenchen. "Intracellular Express: Why Transport Protein Molecules Have Brakes." *Science News* (blog), *ScienceDaily*, October 11, 2010, www.sciencedaily.com/releases/2010/05/100521191233.htm.

Teresi, Dick. "Discover Interview: Lynn Margulis Says She's Not Controversial, She's Right." *Living World* (blog), *Discover*, June 17, 2011, http://discovermagazine.com/2011/apr/16-interview-lynn-margulis-not-controversial-right.

Thomas, Brian. "Circular Arguments Punch Holes in Triceratops Study." *News* (blog), *Institute for Creation Research*, July 21, 2014, https://www.icr.org/article/8218.

———. "Impact Theory of Moon's Origin Fails." *News* (blog), *Institute for Creation Research*, October 28, 2013, https://www.icr.org/article/impact-theory-moons-origin-fails/.

Torley, V. J. "A World-Famous Chemist Tells the Truth: There's *No* Scientist Alive Today Who Understands Macroevolution." *Intelligent Design* (blog), *Uncommon Descent*, March 6, 2014, https://uncommondescent.com/intelligent-design/a-world-famous-chemist-tells-the-truth-theres-no-scientist-alive-today-who-understands-macroevolution/.

Tyson, Peter. "Life's Little Essential." *Evolution* (blog), *NOVA*, January 4, 2014, https://www.pbs.org/wgbh/nova/article/liquid-of-life.

Van Till, Howard J. "Is the Creation a 'Right Stuff' Universe?" *Perspectives on Science and Christian Faith* 54 (1977) 232.

Wald, George. "The Origin of Life." *Scientific American* 191 (August 1954) 44–53.

Warfield, Benjamin Breckinridge. *The Inspiration and Authority of the Bible*. Edited by Samuel G. Craig. Philadelphia: The Presbyterian and Reformed Publishing Co., 1970.

Werner, Carl. *Evolution: The Grand Experiment, Volume 1*. Green Forest, AR: New Leaf Press, 2007.

"What Are the Requirements for Life to Arise and Survive?" *Astrobiology* (blog), *Las Cumbres Observatory*, 2019, https://lco.global/spacebook/what-are-requirements-life-arise-and-survive/.

Wiker, Benjamin. *The Darwin Myth: The Life and Lies of Charles Darwin*. Washington, DC: Regnery, 2009.

Wile, Jay L. "No, It's Not a Tail!" *Archives* (blog), *Proslogion*, May 21, 2012, blog.drwile.com/?p=7621.

Wise, Kurt. "Trilobite Eyes—Ultimate Optics." *Animals* (blog), *Answers in Genesis*, October 1, 2012, https://answersingenesis.org/extinct-animals/trilobite-eyes-ultimate-optics.

## Bibliography

Wood, Bryant. "The Walls of Jericho." *Bible* (blog), *Answers in Genesis*, March 1, 1999, https://answersingenesis.org/archaeology/the-walls-of-jericho.

# Index

*Ambulocetus*, 68, 98–99
Answers In Genesis, 15, 48, 117, 155
apologetics, 51, 124
Avida, 121

Biblical (Great) Flood, 22, 84, 94
Big Bang, 35, 114
billions of years, 18, 23, 74–75, 78, 137
biogeography, 97

catastrophism, 77, 86
Ceolacanthus, 88
change over time, 22, 86, 89
Charles Lyell, 76–77
Christianity, 4, 30, 32, 44, 52, 72
Christopher Hitchens, 23
common ancestor, 63, 68–69, 98, 104, 108
connotations, 21
Coppedge, 6
creation of life, 9, 113
Creation/Evolution Headlines, 153
*Creationist's Trojan Horse*, 123
cytochrome C, 69, 106

Darwin, 62–72, 89
Dawkins, 23, 69, 120, 125
day, 15, 81–82, 150
*Defeating Darwinism by Opening Minds*, 60
Denotations, 21
DNA, 6, 38, 55–57, 106, 122
"junk" DNA, 14, 55–56
doctrine of inspiration, 148
*Drosophila melanogaster*, 65

Edward Blyth, 63, 66
embryos, 103–5
endosymbiotic theory, 114–16
Ernst Haeckle, 103–4
*Ever Since Darwin: Reflections In Natural History*, 56
Evolution Right, 5, 52, 66, 96
Evolution Wrong, 5, 19–20, 48, 55–56, 79, 101, 123

finch beak, 22, 97–98
*Finding Darwin's God: A Scientist's Search for Common Ground between God and Evolution*, 31–37
finite verb form, 81
First Cause, 25
fossil graveyards, 87, 94
fossil record, 48, 85, 89–90, 96
fossils, 78, 83–95

Galileo, 10–11, 133–34
genetic recombination, 118
geologic column, 86
God's Word, 14–15, 41, 81–82, 140–51

homologous molecules, 106–7
homology, 68–69, 100–101

information, 6, 38, 57, 107, 119–22
Institute for Creation Research, 5, 153–54
intelligent design, 37, 53, 119–25, 152
interpretations, 15, 52–53, 81

James Hutton, 74, 76

# Index

Ken Miller, 26, 30–41

language of doubt words, 123
Lyell, 76–77
Lynn Margulis, 26, 70, 116

magic arrows, 113–15
manuscript evidence, 142–45
metaphysics, 50
Michael Ruse, 50
mutations, 22, 28, 30, 34, 64–65, 70, 101, 117–19, 160

natural selection, 26, 46, 63–66, 89
Naturalism, 9–10, 56–57

*On the Origin of Species*, 66
origins, 11, 13, 15, 25–26, 54, 109–18

Pakicetus, 99
Phillip Johnson, 60
philosophy, 25–26, 50–61
planetesimals, 110
polystrate tree trunks, 88
primitive sexual reproduction, 117
prophecy, 30, 139, 145–46

racemic amino acids, 59
radioactive carbon, 77–80

Radioisotopes and the Age of The Earth, 79
rapid subduction, 95
Richard Lewontin, 16–17, 38, 70
RNA, 6, 55, 119
rock dating, 80

science, 9–20, 24–29
Shakespeare, 120
*Signature in the Cell*, 24, 54, 119–20
specified complexity, 122
specified information, 125
Stephen C. Meyer, 24, 54, 119–29
Stephen J. Gould, 6, 10, 53–56
survival of the fittest, 63, 65

talkorigins.org, 28
*The American Biology Teacher*, 74
*The Darwin Myth*, 70
the Goldilocks Syndrome, 116
*Thousands. . .Not Billions*, 80
time, 71–82

uniformitarianism, 76
Urey-Miller experiment, 58

variation, 63–66, 76, 98, 105
vestigial organs, 56, 102–3

www.ingramcontent.com/pod-product-compliance
Lightning Source LLC
Chambersburg PA
CBHW070918180426
**43192CB00038B/1747**